P9-DIW-726

"[A] wake-up call for a brewing crisis. We will not defeat microbes. . . . The reason for concern—reduced [interest from the pharmaceutical companies]—is patently obvious, because you take an antibiotic for a week and a statin for a lifetime. The answer is the need for incentives. . . . This is the only review I know that has the complete story from all perspectives."

—John G. Bartlett, MD
Former Chief of the Division of Infectious Diseases,
Johns Hopkins University School of Medicine

"[P]rovides an insightful overview of the emergence of the most problematic drug-resistant bacteria and the damage and death that result from infections caused by these bacteria. [Spellberg] reminds us of the incredible power of antibiotics and what we stand to lose if antibiotic development does not keep pace with the development of resistance. Most important, he offers innovative solutions to the problem. We need to pay attention."

—John Bradley, MD, FAAP, FIDSA
Director, Division of Infectious Diseases,
Rady Children's Hospital San Diego

rising plague

brad spellberg, md

foreword by
david gilbert, md
past president of the infectious
diseases society of america

rising plague

the **global threat** from **deadly bacteria** and our **dwindling arsenal** to fight them

Prometheus Books
59 John Glenn Drive
Amherst, New York 14228-2119

Published 2009 by Prometheus Books

Inquiries should be addressed to
Prometheus Books
59 John Glenn Drive
Amherst, New York 14228–2119
VOICE: 716–691–0133, ext. 210
FAX: 716–691–0137
WWW.PROMETHEUSBOOKS.COM

13 12 11 10 09 5 4 3 2 1

Library of Congress Cataloging-in-Publication Data

Spellberg, Brad.
Rising plague : the global threat from deadly bacteria and our dwindling arsenal to fight them / by Brad Spellberg.
p. cm.
Includes bibliographical references and index.
ISBN 978–1–59102–750–8 (hardcover : alk. paper)
1. Drug resistance in microorganisms. I Title.

QR177 .S64 2009
616.9'041—dc22

2009019133

Printed in the United States on acid-free paper

contents

foreword

In the spring of 1991, the *Andrea Gail* left Gloucester, Massachusetts, for the fishing grounds of the North Atlantic. Two weeks later and far from port, the unexpected—in fact, unprecedented—happened. Warm air from a low-pressure system coming from one direction, a flow of cool and dry air from a high-pressure system in the opposite direction, AND tropical moisture provided by Hurricane Grace generated skyscraper-sized waves and doomed the *Andrea Gail*. The fishermen could not have predicted the confluence of factors that generated a "perfect storm."[1] *Rising Plague*, by Brad Spellberg, describes a "perfect storm" that is predictable and, in some parts of the country, has already arrived.

For roughly seventy years, we have taken for granted that pneumonia, skin infections, or infections of the bladder and kidney were easily treated with a seemingly endless supply of powerful and safe antibiotics. Unfortunately, this is no longer true. As the patients presented by Dr. Spellberg illustrate, bacteria are increasingly resistant to the antibiotics currently approved by the US Food and Drug Administation (FDA). Concomitantly, for the reasons discussed in this book, the pharmaceutical industry has little interest in investing in the discovery and development of new antibacterials active against the resistant bacteria.

In short, we have a "perfect storm." In contrast to the fishermen on the *Andrea Gail*, we can forecast the future. With no new drugs, physicians are forced to increase the number of prescriptions of currently licensed drugs, which, in turn, will result in more resistance to antibiotics.

The problem can be averted. Dr. Spellberg provides reasonable and pertinent steps of action. However, with so many problems currently facing the nation, it will take effort to make the public, and the legislators who represent us, aware of a looming disaster. A disaster for both the citizens of the United States and the rest of the world. All too often, it takes the death of a celebrity or a massive disease outbreak to motivate human behavior. Dr. Spellberg correctly points out that waiting for tragedy as the supply of effective drugs dwindles and the bacteria become ever more resistant is not an acceptable option. Using his eloquent and easily understood style, Dr. Spellberg calls all of us to action so as to avoid the "rising plague."

David Gilbert, MD
Chief, Infectious Diseases
Providence Portland Medical Center;
Professor of Medicine
Oregon Health & Sciences University;
Past President, Infectious Diseases Society of America (2003);
and Coeditor, *Sanford Guide to Antimicrobial Therapy*

INTRODUCTION

to die from an untreatable infection

no one is safe

"I'm out of antibiotics. She's going to die."

Her doctor looked at me. "Can you help me talk to her husband?"

I paused. Then I glanced back at the computer screen, just to make sure.

Mrs. B. was a pleasant, young homemaker in her mid-twenties. She was happily married, with two small children at home. She had been feeling fine, in her usual state of excellent health. Then, out of the blue, she began to feel tired and weak. She also noticed bruises on her legs and arms. It was when she began bleeding from her gums that she decided to go to the doctor for a checkup. Routine blood tests were sent out. The results were not good.

Mrs. B. had developed leukemia—cancer of the blood. The leukemia, which had erupted initially in her bone marrow, was now rampaging throughout her body. There was no rhyme or reason for it. She hadn't been exposed to any toxins. She hadn't been exposed to radiation. She was just unlucky.

Mrs. B. came to my hospital to receive treatment for her case of leukemia because she had no health insurance, and she could not afford the tens of thousands of dollars it would cost to receive chemotherapy. Harbor-UCLA Medical Center is a county hospital, whose operations are funded by taxpayer dollars. We, and our sister county hospitals and clinics, are the only healthcare refuge for uninsured patients like Mrs. B.

At Harbor-UCLA, Mrs. B. received intensive chemotherapy to try to wipe out the leukemia cells and put her cancer into remission. Unfortunately, and inevitably, the combination of the leukemia and the chemotherapy also wiped out her normal white blood cells, which were supposed to be defending her body from infection.

When she spiked her first fever, Mrs. B.'s internal medicine doctors called in the infectious-diseases consultants—that is, my teammates and me—to advise on appropriate treatment. When I first met Mrs. B., she was in "isolation," which means we were taking special precautions to try to keep her from being exposed to bacteria, fungi, and anything else that could infect her. The door to her room was kept closed, and a sign reminded all visitors to wash their hands. I knocked and opened the door.

The room was still. Outside, that famous, mid-morning Los Angeles sun was shining, but the curtains in front of the window were drawn. The overhead lights were off, and the dim room was lit only by the ambient sunlight creeping around the corners of the curtains. In the middle of the room, Mrs. B. lay in bed, resting quietly, her eyes closed. Her husband sat in a chair at the side of the bed, and their two toddlers were on his lap. Even the children were somber. They sat in silence, just watching her breathe.

Mrs. B. turned her head slightly, and her family looked over at me as I walked through the door and introduced myself.

"How are you feeling?" I asked as I washed my hands.

Mrs. B. tried to speak but her words were garbled, her mouth dry and sore from the chemotherapy. She cleared her throat and swallowed. "Not so good. I feel very weak." Her voice cracked again, as if it was difficult to talk. Her face was drawn and pale, and her lips chapped. Her hair was matted with sweat.

"Yes, the weakness is from the leukemia," I explained. "I'm from the infection team, and I understand you've been having fevers. That's why your doctors called and asked me to come see you." I stepped toward the bed. "Do you mind if I examine your wife for a few minutes?" I asked Mrs. B.'s husband.

He looked me over for a minute. Then, without a word he nodded, rose, and led their kids outside, closing the door behind them.

I drew the curtain around the bed. "Do you have any pain anywhere?"

She shook her head.

I carefully examined Mrs. B. to try to find the source of her fever. She was very pale from the anemia that was caused by her leukemia. She had bruises up and down her arms, the inevitable marks of a hospitalized patient, caused by incessant blood draws ordered by her doctors. Mrs. B.'s mouth was filled with superficial ulcerations, some of which looked mildly inflamed. These sores were typical side effects from the chemotherapy she was receiving for the leukemia. My examination revealed nothing that could lead me to the source of Mrs. B.'s fevers.

"Well, things look okay right now," I told her. "But we still need to try to find out where these fevers are coming from so we can make sure to give you the right antibiotics. In the meantime we are going to put you on very powerful antibiotics to try to protect you from as many different bacteria as possible."

She nodded.

"Do you have any questions?"

She thought for a few minutes and then asked, "Do you know how long I will have to be here?"

"We need to get your white blood cells back up. That's the most important thing. It's going to be a few weeks more at least."

Mrs. B. nodded and exhaled, her exhaustion palpable.

As I stepped outside, and sent her family back in to be with her, I considered the facts. I had no idea what type of infection Mrs. B. might have. Her lab tests were all negative. Her physical exam had not been revealing. But the one thing I knew was that the leukemia had massively weakened Mrs. B.'s immune system, and chemotherapy had finished the job. So Mrs. B. had no functioning immune system to fend off infections. Now that she was infected somewhere in her body, she had a high risk of dying unless effective treatment was initiated immediately. Knowing this, I started "big gun" antibiotics, designed to treat virtually any bacterial infection that Mrs. B. might have.

On those powerful antibiotics, Mrs. B.'s fever went away, and she began to feel better over the next few days.

When Mrs. B. spiked her second fever a week later despite already being on powerful antibiotics, I was very worried. I tried adding new antibiotics, but this time her fever did not go away. Mrs. B. would need to survive for several more weeks before her immune system had any chance of recovering from the chemotherapy. Until then, antibiotics were her only hope to stay alive.

By now, Mrs. B. had developed a cough. Mrs. B.'s chest X-ray showed that she had developed pneumonia (an infection in her lungs). The antibiotics that we were giving her should have been able to treat the pneumonia, but for some reason they were not working. Day after day Mrs. B. continued to spike fevers despite the antibiotics, and her cough got worse. Then her heart rate shot up. Her husband, afraid and frustrated by our inability to make his young wife better, never left her side.

You could see the terror in her face as Mrs. B. began to gasp for air, her oxygen level falling dangerously low because her lungs were failing from the pneumonia. Her lungs got so bad that we had to put her on a breathing machine—a mechanical ventilator—to keep oxygen flowing into her lungs and carbon dioxide flowing out. Her chest X-ray showed that her pneumonia was expanding, getting worse day by day despite antibiotics. We sent her sputum (the thick phlegm she was coughing up) and her blood to the microbiology laboratory for culture, hoping to identify the exact bacterium causing the infection.

Two days later, the lab posted results.

Acinetobacter baumanii. Or, as I call it, "Big, bad mojo." It was, and still is, one of the most antibiotic-resistant bacteria around.

"It's resistant to almost every antibiotic," I told Mrs. B.'s internal medicine doctor, as I scanned down the laboratory report on the computer. I had no choice. According to the laboratory, there was only one antibiotic left on the planet that could be used to treat the *Acinetobacter*. And this antibiotic was the last line of defense against life-threatening, hospital-acquired infections: imipenem. I immediately changed her antibiotics, and started Mrs. B. on the imipenem.

Initially Mrs. B. seemed to improve on the new antibiotic. Her fever began to creep down. Her breathing even got better, although she remained on the mechanical ventilator.

Several days later, however, her fever worsened once again. And her lungs got worse again. Then her blood pressure started to fall. Her kidneys began to fail. We sent out new cultures.

I double-checked the computer, just to be sure I had read it right. "It's resistant to everything now," I confirmed. The *Acinetobacter* had developed resistance to the imipenem even as we were treating Mrs. B. with it—the bacterium had learned how to defeat our most powerful antibiotic in the span of a week.

"Her husband is angry," Mrs. B.'s internal medicine resident[1] told me. "He can't understand why she's not getting better, and he thinks we are not doing everything we can. I tried to explain to him, but he just wasn't getting it. Can you talk to him for me?"

"Okay. Let's go."

So the resident led me to Mrs. B.'s bedside. Mrs. B. was floridly septic—that is, her infection had spread throughout her entire body and the systemic inflammation was causing all of her blood vessels to open so widely that her blood pressure plummeted. We had to pour liters of fluid and medications into her to fight the sepsis, to try to keep her blood pressure up, and keep her alive. Of course, in the end, if we couldn't kill the bacteria causing the sepsis, no amount of fluid or other medications was going to be able to keep her blood pressure up.

Mrs. B. lay in a twenty-first-century intensive care unit, receiving maximal medical therapy. She had a tube down her throat so our mechanical ventilator could breathe for her. She had a tube winding down through her nose and into her stomach so we could administer ultranutritious liquid food. She had a tube in her bladder to collect her urine. Since she couldn't get out of bed to go to the bathroom, she was wearing a diaper to collect her stool. The only natural orifices into which we had not jammed plastic tubes or covered with a diaper were her ears, which were instead covered by the tape wrapped around her head to keep her breathing tube in place.

Since the doctors had run out of natural orifices into which to stick things, we had torn new orifices into her body, placing IV tubes into the peripheral veins in her arms and legs, and jamming a gigantic plastic catheter into the jugular vein in her neck. We were pumping liters of saline into her body, and infusing powerful "pressor" medications, all to keep her blood pressure from plummeting due to her severe sepsis. We were administering the most powerful antibiotics that existed anywhere in the world.

Despite all of this, there she lay, rotting away from within.

Her husband was waiting in the hallway. We went out to talk to him. He was pacing, his hands clenched, his jaw muscles pulsing. When we came into the hall, he turned to face us with a grim countenance.

The resident tried to explain that Mr. B.'s wife had an infection that was not responding to treatment.

"Try something else," Mr. B. insisted. "I want everything done."

The resident looked at me.

I stepped forward. I reintroduced myself to Mrs. B.'s husband as an infection specialist. Mr. B. did not seem impressed. Then I told him something that I had never said before. "I've run out of antibiotics. I have nothing left to use."

"Try something else!"

"Sir, I am very, very sorry. But we have done everything we can. This infection is resistant to every antibiotic there is. There is nothing else to try."

Mrs. B. died the day after I spoke with her husband, despite the maximum effort of numerous doctors and nurses, and despite cutting-edge medical technology. She left behind a devastated widower and two small children. That was several years ago. While such infections had been well described in the medical literature, it was the first time I had personally ever seen a pan-resistant infection—that is, an infection caused by a microbe resistant to every antibiotic option. But it was not the last time.

Descendents of the same bacterium that killed Mrs. B. still plague my hospital and thousands of other hospitals throughout the United States and the world.[2] The same type of bacterium, *Acinetobacter*, has maimed many injured soldiers from Iraq and Afghanistan.[3] It has caused infections in Intensive Care Units across the globe, and there is no sign that its spread is slowing down. To the contrary, like so many other types of deadly infections, its spread is clearly speeding up.[4]

What we need are new antibiotics to allow us to treat these infections. Unfortunately, at the same time the bacteria are getting more resistant, fewer and fewer new antibiotics are being developed.[5] The microbes have leapt ahead, and medicine is not even trying to keep pace. How can that be? Why is it that many pharmaceutical companies are no longer interested in discovering and developing new antibiotics? How is it that the microbes have moved so far, so fast to become resistant to our current antibiotics? And how is that our society seems totally disinterested in keeping pace with the microbes, thereby preventing what could become a widespread catastrophe?

That's what this book is about, the rising perfect storm of drug-resistant infectious diseases and the lack of new antibiotics with which to treat them. My goals are threefold: (1) to inform you of the scope of the convergent problems; (2) to explain their causes; and (3) to advocate solutions that experts who have studied these problems have proposed but that have gone unheeded for years.

Think antibiotic-resistant infections will never affect you? So did Mrs. B. So do most people that die from them.

CHAPTER 1
hard lessons learned from mrs. c.

MRS. C. AND THE CASE OF THE MYSTERIOUS FEVER

I first met Mrs. C. before I became an infectious-diseases specialist. At the time, I was a senior resident physician, in my final year of training in internal medicine. My team and I were taking "long call" that night, meaning that I, along with some interns working under my supervision, was responsible for admitting patients to the hospital overnight, from 2 p.m. to 8 a.m. the following morning.

The emergency room physician called me in the early afternoon to admit Mrs. C., who had come to the emergency room complaining of generalized body aches. Mrs. C. was a very pleasant, fiftyish mom with a large extended family. Her husband was standing next to her gurney in the ER when I first met them.

"Hi, Doctor," Mrs. C. greeted me after I introduced myself. She had a matronly air about her and exuded a friendly, hospitable aura.

"So, what brings you in today?" I asked.

"I just haven't been feeling well recently. I'm achy and tired." She frowned and shook her head.

"Hmm. How long has this been going on?"

"I guess about a week or so. Maybe a little more."

"And when you say *achy*, where do you have pain?"

"Mostly in my joints. Mostly my elbows and my knees, and a little in my hands. And I just have muscle aches everywhere. My back hurts, and my arms. I just feel bad overall."

"Have you had any fevers?"

Mrs. C. nodded. "Yes, I think so. I've been having fevers almost daily for a week."

"She was one hundred and two yesterday," her husband said. "We checked it." He was a burly gentleman with a broad face, a salt-of-the-earth type.

"Okay. When you say you've been feeling tired, is it more a sleepiness, or is it more that you feel weak?"

"It's more weakness than sleepiness."

"Are you able to go to work, or walk around the house?" I asked. "How much activity can you do?"

"It's pretty bad. I haven't gone to work since last week, both from the weakness and the joint and muscle aches. I'm really not getting around the house much at all."

"Have you had any rash?"

"No, I don't think so." She looked at her husband. He shook his head. "Not that I've seen."

"Do you have any history of arthritis, or immune diseases like lupus?"

"No."

"Does anyone in the family have any history of arthritis, lupus, or any other serious medical conditions?"

"Well, both my parents had diabetes and my father had a heart attack."

"No arthritis?"

"Not that I know of."

"How about thyroid disease?"

"No."

"And you don't have diabetes?"

"I've been checked before but the tests said I was normal. I do have high blood pressure."

"Okay." I paused to consider. "Have you been out of the country recently, or gone on any trips?"

"I was in Mexico a few years ago. But nothing out of the country since then."

"And you didn't get sick while in Mexico?"

"No, not that I recall."

"How long were you there?"

"Oh, let me see. Probably about two weeks—right, honey?"

Her husband nodded. "That's about right."

"Any recent travel out of Los Angeles?"

"We went to Las Vegas last year for a few days."

"That's it?"

"That's it. We're homebodies more or less."

"Fair enough. Do you have any pets?"

"Two dogs."

"Okay. Do you take any medications?"

She named a medicine for high blood pressure, which had been her only medical problem prior to the past week.

"Have you taken any herbal medications or supplements, or any other over-the-counter drugs recently."

"Nope."

"So, nothing new in your life, nothing different in the last month or so? No recent colds or coughs or the flu?"

"No, I've pretty much been fine until just the last week or so."

Her husband shifted.

I watched as he frowned at Mrs. C.

She looked back up at him.

"Okay, what am I missing?" I asked.

"You should tell him about the vein," Mr. C. said.

"Oh, it's probably nothing," Mrs. C. said.

"Tell him," Mr. C. insisted.

I raised my eyebrows. "What's this about a vein?"

Mrs. C. shrugged. "Well, I don't think this has anything to do with it, but before all these symptoms started I popped a varicose vein with a safety pin."

"Excuse me?" I asked.

She offered a sheepish grin. "Well, I have this ugly varicose vein on my leg. So I took a safety pin and tried to pop the vein."

"You put the safety pin into the varicose vein?" I asked.

"Well, first I boiled the pin to sterilize it. Then I just poked a hole into the vein."

"And what happened?"

"Well, some blood came out and I put pressure on it and it stopped bleeding after a few minutes."

"And how many days before your symptoms started did you do this?"

She looked at her husband. "Oh, I don't know, maybe a week or ten days?"

He nodded. "Something like a week or two. I don't know if it has anything to do with anything, but I thought she should tell you."

I shrugged. "I don't know either, but the more information the better."

I asked a few more questions and found out nothing additional of significance. I then examined Mrs. C. from head to toe. I didn't find any evidence of joint swelling. She did not have any rash. Her heart and lung examinations were normal. She had tenderness when I pressed on her back muscles and on her leg muscles, but she had normal function in her nerves. She was a little weak in her arms and legs, but that seemed to be due to pain with movement more than anything. Her varicose vein was still present, but didn't seem to be inflamed. In other words, by the time I was done examining Mrs. C., I still had no idea what was happening with her.

Mrs. C.'s vital signs, recorded by the emergency room nurse, showed a very mild temperature elevation, barely into the true fever range. Her heart rate and blood pressure were mildly elevated. It really was not a very impressive presentation. She just didn't seem very sick. I was thinking Mrs. C. probably had a mild viral syndrome, or perhaps early rheumatoid arthritis or thyroid disease. I made a list of possible causes for her symptoms and sent off a battery of tests, checking for infection, arthritis, thyroid disease, and some other possibilities. I assigned Mrs. C. to an intern working with me, and asked him to go meet Mrs. C. and repeat the assessment. Then I moved on. I had more patients to see. It was going to be a busy night.

It was indeed a hard call night. My teammates and I were up much of the evening admitting new patients and caring for those patients of ours already in the hospital. I gratefully dropped into bed in the on-call room at around three in the morning. I fell fast asleep. I got beeped again a while later, and admitted my final patient at around five in the morning. Then I trudged back up to the on-call room to try to catch another hour or two of sleep. At about six thirty in the morning I got beeped yet again. I groaned and rolled over to check my pager. It was a phone number on the wards, and there was a code behind it that identified the person paging me as an intern.

I exhaled and gathered myself. Then I reached over, grabbed the phone on the wall, and called the number.

The intern, whom we'll call Peter, answered. "Sorry to bother you," Peter said, "but [Mrs. C.] is not looking good at all."

"Really?"

"All of a sudden, she's really doing poorly. She's confused and thrashing about."

I frowned. "How are her vital signs?"

"She febrile, tachycardic, and tachypneic," he said, meaning that she had a fever, a rapid heart rate, and was breathing rapidly.

"How's her blood pressure?"

"It's holding."

"Okay, I'll be right down."

Overnight Mrs. C. had been moved from the emergency room to a hospital bed upstairs. I climbed out of bed, rumpled and bleary-eyed, and took the elevator down to her ward. As I walked through the hallway toward her room, I saw her family gathered around outside. Mrs. C.'s siblings, children, and nieces and nephews were gathered there, along with her husband. They were talking in hushed, worried voices, and their conversations fell silent instantly as they noticed me walking toward them. Mrs. C.'s daughter approached me.

"What's going on with my mother?"

"I'm not sure. Give me a few minutes to examine her. I'll come talk to you guys as soon as I've finished, okay?"

I was keenly aware that the entire family was gazing fixedly upon me as I walked into Mrs. C.'s room. The intern was at her bedside. The first thing that was obvious upon my arrival was that this was a totally different Mrs. C. than the woman I had met just fourteen or fifteen hours earlier. She was completely confused. The thrashing about in her bed was actually a rhythmic snapping movement of her right arm and a jerking of her neck. Her eyes were glazed over. Because she had not appeared severely ill initially, she was in a regular hospital ward bed, rather than in an intensive care unit bed. So she was not hooked up to devices for continuous monitoring of her vital signs. Rather, the ward nurse was checking her vital signs manually once every four to eight hours.

"She looks like she's having seizures," I told the intern. "We need some vital signs. Grab the nurse's flow sheets and bring them back so I can see what her most recent vital signs were." He hurried off.

"Mrs. C.?" I called out, and put my hand on her shoulder. She continued her spasms, with no sign of recognition. Her eyes were unfocused and her breathing erratic. I did a quick examination, listening to her heart and lungs with my stethoscope and pressing on her abdomen to see if it was tender. But I found nothing new since the prior afternoon.

Peter brought back the vital sign sheet. As he had told me, the vital sign sheet showed a steady increase in Mrs. C.'s heart rate and breathing rate overnight. Thus far her blood pressure had not begun to fall, but Mrs. C. had obviously become very sick.

We rushed to grab Mrs. C.'s nurse, who was standing right outside the room. Fortunately, Mrs. C.'s nurse was one of the hospital stalwarts. We'll call her Jacqueline. Jacqueline didn't put up with nonsense from patients, doctors, pharmacists, or anyone else. She knew her business, and she was worried about Mrs. C.

"Mrs. C. looks terrible," I said.

Jacqueline looked at me as if I was Captain Obvious. "I know. That's why I paged the intern."

"How long has she been like this?"

"She's been getting worse over the last hour or so. And then even worse in the last fifteen to twenty minutes. That seizure movement started just a few minutes ago."

I considered for a moment. Then I asked Jacqueline to fill a syringe with a medication to break the seizure, which she rushed off to do. When she brought me the syringe, we took a special, portable vital sign monitor into the room with us. Jacqueline attached the monitor to Mrs. C.'s finger, and attached a blood pressure cuff to her arm. The monitor would tell us Mrs. C.'s heart rate and would also give us a read of the oxygen level in Mrs. C.'s blood.

I grabbed Mrs. C.'s IV, attached the syringe, and began slowly pushing the medication into the IV. At first, nothing happened. But, halfway through the syringe, Mrs. C. began to relax a little bit. Then her spasms stopped. She lay back in bed.

"Yes!" I thought. Things were going our way. Then I finished injecting the medication and detached the syringe from the IV. We watched Mrs. C. for a few moments.

"Looks like we broke the seizure. Cycle her blood pressure please," I said.

Jacqueline pressed the button to activate the automatic blood pressure cuff. The device beeped and the cuff began filling with air. Then the inward air flow stopped and slowly the air came out of the cuff. The cuff monitor ticked with each release of air. A few agonizing moments passed, as the cuff released pressure bit by bit.

Finally the blood pressure monitor flashed a reading: "100/50."

"Uh oh. The BP is falling, and it's way down from yesterday afternoon," I said to Peter. "Jacqueline, let's give her a liter of saline." Jacqueline had already prepared the saline and began to hang the bag up on the IV pole.

"What is going on with her?" I wondered. I was wracking my brain, trying to figure out what was happening and what could be done to reverse it.

As we stood at the bedside, Mrs. C.'s breathing began to slow down right before our eyes. Suddenly she took on a "fish out of water" appearance, with pursed lips sucking in great gasps of air. But between each of her great breaths, an agonizing five to ten seconds passed.

The oxygen level monitor attached to Mrs. C.'s finger began alarming, indicating that the level of oxygen in her blood was falling. I reached for the oxygen mask hanging behind Mrs. C.'s bed. "Get me a bag right now, and call anesthesia STAT please," I called to Jacqueline. "We're intubating this patient!"

Jacqueline hurried outside to carry out the order and to summon more nurses to help.

"Keep your hand on her wrist," I told Peter, as I slipped an oxygen mask over Mrs. C.'s face. "Watch the pulse."

I heard the page operator calling the on-call anesthesiologist overhead. Jacqueline hurried back with an "Ambu" bag. The facemask I had put on Mrs. C. blew oxygen into her face, but the air would only get down into her lungs if Mrs. C. initiated a breath. So I replaced the oxygen face mask with the face mask attached to the Ambu bag and began squeezing the bag. The Ambu bag forced air down into Mrs. C.'s lungs, helping her to breathe even if she couldn't.

"Wait!" my intern shouted. "We're losing the pulse!" He frantically grabbed her wrist and shifted his fingers. "No pulse!"

I put my fingers on her carotid artery. No pulse. Uh oh!

"Call a code!"

An army of nurses had gathered in the room and one of them rushed out to call the code. Mrs. C.'s family was looking on beyond the nurses. I heard a shriek from them.

"Get a backboard underneath her!" I said.

Peter and I tilted Mrs. C. up on her side and Jacqueline slipped a backboard beneath her. I started chest compressions, looking down at this pulseless, nonbreathing woman who had seemed so alive and not very sick just the prior afternoon. "Keep squeezing the Ambu bag," I instructed Peter.

A group of nurses rushed into the room with the code cart and began opening medication boxes.

"Get that saline bolus running wide open please!" I called out as I continued chest compressions. "And push a milligram of epinephrine."

The Code Blue call rang over the loudspeaker as the page operator announced the code. Within a few moments, as I continued chest compressions, the overhead page brought an army of doctors to help. Amid the blitzkrieg of activity, someone connected Mrs. C. to an electrocardiogram so we could determine her heart rhythm.

"Looks like v-fib," one of the doctors said, meaning ventricular fibriliation, a fatal arrhythmia. "Let's shock her." He reached back to grab the paddles.

"Wait, she doesn't have pads on!" I huffed, as I continued compressions. "Get some pads."

One of the nurses opened a package of two conducting pads and charged forward. I stopped compressions and stood back. The nurse slapped the adhesive conducting pads on Mrs. C.'s chest.

"Clear!"

Mrs. C.'s body jolted violently in bed as the paddles discharged.

Despite the conducting pads, the smell of burning flesh filled the room.

"Check pulse, please," I shouted.

"No pulse."

The electrocardiogram still read fibrillation.

"Three hundred joules. Clear!"

Mrs. C. spasmed once again, the bed shaking from the jolt, as three hundred joules of electricity flowed from the paddles, through the conducting pads, through her chest wall, and into her heart muscle.

"Check pulse."

Another agonizing moment passed as Peter checked Mrs. C.'s pulse.

"Her pulse is back!" he shouted.

"Check again."

"Yep, it's definitely there."

I checked her carotid artery and concurred. I was sweating bullets now, and the adrenaline was pumping. I breathed a massive sigh of relief.

Now that her pulse was back, the anesthesiologist immediately intubated Mrs. C. and connected her breathing tube to a mechanical ventilator. The other doctors and nurses who arrived helped place new IV in her so we could give even more medications. We pushed her bed into the intensive care unit. It all happened in fifteen minutes.

And I still had no idea why poor Mrs. C. had gotten so sick, so fast.

Mrs. C.'s family was distraught. As we wheeled Mrs. C. out of her room and toward the ICU, I saw them staggering about in the hall. Half were crying and half were gazing about, faces glazed with shock.

Up in the ICU, I explained what little I knew to the critical care team, and we came up with a plan to support Mrs. C.'s blood pressure, start broad-spectrum antibiotics, and send off other diagnostic tests. A few minutes later, after we had gotten Mrs. C. stabilized in the ICU, I knew it was time to face the music. The critical care team would handle Mrs. C.'s immediate medical needs. Meanwhile, I trudged back to meet the family.

I was completely spent. And I had trepidation about how the family might react to seeing their mother "code" right before their eyes. I was feeling guilty that Mrs. C. had gotten so sick while under my care. I was racking my brain, trying to think if I had missed something. Had some clue or some answer eluded me?

I asked Mrs. C.'s family to follow me into a family waiting room, where we could talk in private. I sat in the middle of the room, and they filed in after me, surrounding me with eyes searching for answers. Mrs. C.'s husband and children sat directly in front of me, looking me right in the eye.

"Did she pass?" her husband whispered, hands clenched, knuckles blanched.

"No, no, we got her back. She's fighting and she's still with us."

They veritably exploded in relief, in a collective exhalation. Tears broke through and welled at Mr. C.'s eyes. I had to fight off my own tears.

"I can't promise you what's going to happen from here on out, but for now, she seems to have stabilized," I told them. I explained that we still did not know what was going on, but that we were going to do everything in our power to figure it out and to treat her with the most powerful medications we had. They had questions. I had few answers, but I tried to reassure them that we were doing everything we could.

I had been worried about how the family would react. But Mr. C.—whom I had met the first moment I met Mrs. C. in the emergency room—grabbed my hand, looked me in the eye, and said, "Thank you for saving her."

He thought he had seen her die right before his eyes. He and his children had thought she was gone. More than anything, they were grateful that she was still alive.

In the aftermath of the family meeting, I walked to a quiet corner and collapsed into a chair. I put my hands over my face and rubbed my eyes. I just needed a few moments to gather myself. What could have happened? What could be going on with Mrs. C.?

MYSTERY SOLVED!

The answer came from an unexpected source. Because she had been febrile, and because infection was on the list of possible causes of her symptoms, we had drawn blood in the emergency room to send to the laboratory for bacterial cultures. I never really thought Mrs. C. had bacteria in her blood. She had looked far too healthy for that. But we checked the test just to dot the i's and cross the t's. And lo and behold, that very morning, her blood cultures grew bacteria. And the bacteria turned out to be methicillin-resistant *Staphylococcus aureus*, an antibiotic-resistant, highly virulent organism. I was shocked.

I was shocked all the more when we performed a computerized tomography scan (colloquially known as a "CAT scan") of her back, and found that Mrs. C. had no fewer than a dozen separate abscesses in the muscles running up and down her back. Every muscle was affected. The abscesses were coalescing around her spine. Never in my wildest dreams would I have imagined such a severe case of staph infection in a patient who had looked so well on presentation to the hospital.

Thankfully, Mrs. C. survived. We treated her aggressively with a combination of last-ditch antibiotics and surgery. At the time, there was only one antibiotic—vancomycin—that really could be used to treat her infection. Mrs. C. was in the hospital for many weeks. She had many touch-and-go moments. And despite our best efforts and our most powerful antibiotic, she became partially paralyzed in her legs as the bacteria inexorably spread toward her spinal cord. But at least she was alive. Eventually we managed to halt the bacterial spread and save much of the strength in her legs.

Was Mrs. C.'s infection caused by the "popping" of the varicose vein by a safety pin? We'll never know for sure, but it does seem to be the most likely explanation for how the bacteria got into her bloodstream. Such an innocent act, and with a sterilized needle no less, to result in such a horrible outcome. But the key is, the bacterium, *Staphylococcus aureus*, lives on our skin. So sterilizing the needle would not prevent the needle from forcing the bacteria on the skin into the blood as the needle passed through the skin.

I was not the only one baffled by her case. Mrs. C.'s case was very puzzling even to the subspecialists. The intensive-care doctors, the infectious-diseases consultants, and the surgeons who helped care for her abscesses were equally surprised by the severity of disease given her benign presentation to the hospital. After a prolonged and extremely expensive hospitalization (covered by the taxpayers of Los Angeles), Mrs. C. was eventually discharged from the hospital. She was left with severe weakness in both of her legs, preventing her from walking or even standing upright on her own. From the hospital, she was transferred straight to an inpatient rehabilitation center, where she would need to spend months in an intensive program to try to reverse some of her weakness. Unfortunately Mrs. C. had some degree of damage to her spine from which she would never fully recover, which would result in permanent weakness and disability. Keep Mrs. C. in your thoughts as you read further. Her case is instructive in many ways.

CHAPTER 2
infections, antibiotics, and antibiotic resistance

INFECTIONS AS A CAUSE OF DEATH

When my colleagues or I give interviews to the press on the rise of antibiotic-resistant microbes, often the first question asked by bemused reporters is, "Do people really still die from infections?"

This question never ceases to amaze me, because as an infectious-diseases specialist, I am painfully aware of how many patients die from infections. But until quite recently, the public perception mainly had been that infectious diseases occur in third-world countries, not in the United States or other countries with advanced medical technology. After all, hasn't the combination of public health, vaccines, and antibiotics completely wiped infections off the map?

In the wake of the astonishing success of antibiotics, in 1969, the US surgeon general, Dr. William Stewart, is said to have declared, "It is time to close the book on Infectious Diseases and declare the war against pestilence won."[1] Although this statement appears to be apocryphal,[2] it clearly reflects a long-standing general sentiment among the medical and scientific communities, as well as the public.[3] Indeed, as early as 1951—a decade after the first successful use of penicillin in humans—Sir Frank

MacFarlane Burnet, one of the world's leading experts on the immune system, wrote, "With full use of the knowledge we already possess, the effective control of every important infectious disease . . . is possible."[4] His opinion on this matter had not changed even eleven years later, in 1962,[5] which was two years after he won the Nobel Prize in Medicine for his fundamental research into the immune system. Jerry Pier, professor of medicine, microbiology, and molecular genetics at Harvard Medical School, has published a pithy letter about Burnet's beliefs titled "On the Greatly Exaggerated Reports of the Death of Infectious Diseases," in which Dr. Pier commented, "I find it personally comforting such statements are not an impediment to winning a Nobel Prize or being recognized for making major contributions to science and medicine."[6]

These erroneous beliefs about the impending conquest of infectious diseases due to the success of antibiotics were no passing fancy—they had staying power. A quarter-century after Burnet first declared infections conquered, one of the leading physicians in the country, Dr. Robert Petersdorf, who was himself an infectious-diseases specialist, espoused the idea that the United States was training far more new infectious-diseases specialists than were needed. In the late 1970s he actually wrote, "I cannot conceive of the need for . . . more Infectious Disease specialists . . . unless they spend their time culturing each other."[7] Even as late as 1985, Dr. Petersdorf's opinion had not changed. That year he gave a keynote address at the annual meeting of the Infectious Diseases Society of America, and even amid that bastion of the infectious-diseases medical world, challenged his colleagues by saying, "The millennium where fellows in Infectious Disease will culture one another is almost here."[8]

This prevalent belief that infectious diseases had been overcome by modern medical technology should be remembered as one of the greatest blunders in the history of biomedical sciences. More than a half century after Burnet's declaration of the demise of infections as a relevant problem to civilization, we can now ask, have we, in fact, won the war against infectious diseases? As of 2002, which is the latest year data are available from the World Health Organization (WHO), infections were the second leading cause of death worldwide, killing nearly fifteen million people, which was almost one in three deaths across the globe.[9] By one estimate, 170,000 Americans died of infections in 1996, making them the third leading cause of death in the United States, and representing a doubling in the number of deaths due to infections over the previous three decades.[10] But that figure likely represents a gross underestimate of the true number of infection-related deaths per year in the United States in the twenty-first century. Just the top three infectious causes of death in the United States (sepsis, influenza, and pneumonia) kill an estimated 250,000 to 300,000 Americans per year.[11] Particularly in light of the continued increase in the frequency of antibiotic resistance over the last decade, the current number of Americans that die of infections is likely to be in excess of 300,000 per year. That same annual death rate (about 1 per 1,000 people in the country) is reflective of the infection-related death rates in all countries with advanced medical technology, such as Western Europe, parts of Asia including Japan and Taiwan, and Australia. In other parts of the world, where medical and public-health infrastructures are less well developed, the death rate from infection is much higher.

Some infections, such as pneumonia (lung infection), meningitis (brain infection), and cellulitis (skin infection), are commonly acquired by otherwise healthy people going about their routine daily life. Other infections, called "nosocomial infections," are acquired by people who are already in the hospital. These hospital-acquired infections have become a modern plague, relentlessly attacking countries with advanced medical technology. Nosocomial infections are, in fact, a side effect of modern intensive care, and of all the tools, techniques, and medicines we have developed over the past century to prolong the lives of sick, hospitalized patients.

There is a perception in the public, and even among some physicians, that hospital-acquired infections occur only in "dirty" or "sloppy" hospitals. Balderdash! Hospital-acquired infections are the natural by-product of concentrating sick patients in hospitals and exposing those sick patients to all of the invasive procedures modern physicians use to keep patients alive for as long as possible. For example, sticking a plastic tube (a "catheter") through the skin and into a vein is essential to modern medical care, as it allows doctors to administer lifesaving medications.

But placing catheters through the skin and into veins also provides an artificial "highway" for the bacteria to invade our bodies, enabling them to bypass the virtually impenetrable outer wall of skin that normally protects us from them. Furthermore, the plastic surface of the catheters provides shelter for bacteria to stick to and to hide from our immune systems. Similarly, sticking a tube down the throat of someone who can't breathe so a mechanical ventilator can breathe for them is a lifesaving cornerstone of intensive care. But as a side effect, it allows contents from the mouth, esophagus, and stomach, which are rich in bacteria, to trickle down into the airway, massively increasing the risk of pneumonia. Chemotherapy is how doctors treat lethal cancers, but as a side effect, it wipes out patients' immune systems, predisposing them to infection. Surgeries are often lifesaving, but they also disrupt normal anatomical barriers to infection. Administration of nutrition through the vein is critical to keep patients alive when their intestines aren't working, but creation of such a rich environment of nutrients in the blood enables bacteria or fungi to thrive in the blood. Each of these facets of modern medical care saves lives, but, as a side effect, each allows bacteria and other microbes to enter the body via routes that normally are not available to them, so they can thrive inside of our tissues.

So, hospital-acquired infections are *not* necessarily a sign of a sloppy or poorly run hospital. Rather, they are by and large the outcome of twenty-first-century medicine being practiced to the maximum. While we may be able to reduce the frequency of these infections, no amount of bleach and no amount of obsessive-compulsive hand washing is going to completely eliminate them. Hospital-acquired infections have been with us ever since there were hospitals, and they will continue to be with us as long as hospitals exist.

According to the US Centers for Disease Control and Prevention (CDC), an astonishing 1.7 million Americans in the year 2002 acquired infections in hospitals after being hospitalized for another reason.[12] An appalling 99,000 of those patients died of their infections. Most of these infections were caused by antibiotic-resistant bacteria. To reiterate, these patients came into the hospital with a heart attack, or cancer, or trauma after a car accident, or to have elective surgery, or with some other medical problem, and then ended up dying of infection that they picked up in the hospital. Data are not available after 2002. Given the aging of the US population, the increasingly intensive care being administered in hospitals, and the increasing antibiotic resistance that is exploding through hospitals, the number of people who die from hospital-acquired infections is unquestionably much higher now, and is almost certainly more than 100,000 per year in the United States alone.

It is also generally underappreciated that infections are a major cause of death in patients with cancer. Many cancer deaths are actually caused by infections that occur because the patients' immune systems have been weakened by the cancer itself or by the chemotherapy used to treat the cancer. For example, infectious sepsis alone kills more than 45,000 cancer patients per year in the United States.[13] It is conceivable that were these deaths counted as infectious deaths, infectious diseases would surpass cancer as the second leading cause of death in the United States.

But we should not think of infections, and even antibiotic-resistant infections, as being exclusively related to sick people in hospitals. As Mrs. C. quite clearly demonstrated, and as we shall see over and over again, in the twenty-first century it has become routine for healthy people going about their normal daily lives to acquire and develop severe complications from—or even die from—antibiotic-resistant infections.

Finally, when considering that infectious diseases are the second-most common cause of death in the world, and third-most common in the United States, we must remember that this incidence of infectious-related death occurs despite currently available antibiotics. Throughout much of the history of human civilization, infectious diseases have been among the most frequent killers of humans.[14] The only reason that cardiovascular disease and cancer exceed infections as a cause of death in the modern era is that fewer people now die of infections at a young age. So people are living long enough to die of heart attacks and breast, colon, or other cancers in large part because public-health measures, vaccines, and antibiotics keep them alive that long. Were we to lose most or all effective antibiotics to the development of resistance, the list of the leading causes of death across the globe would undoubtedly be reshuffled to be once again led by infectious diseases.

THE ANTIBIOTIC REVOLUTION

From the earliest days of both Western and Eastern medical tradition, physicians have tried to find "magic bullets" to cure diseases. Indeed, for nearly three thousand years, physicians in China, Egypt, and Europe (beginning with the Greeks) experimented with natural substances as curative drugs for inflammatory disorders and wounds. The ancient Greeks used natural substances such as ground onion, woods like myrrh, and wine and honey to treat or prevent wound infections.[15] They also used caustic substances, such as chalcocite (a mineral form of copper sulfide), galena (a mineral form of lead sulfide), and alum (potassium aluminum sulfate) to prevent or treat wound infections. These caustic substances were most likely effective at preventing or treating infections by actually causing minor injury to the tissue around the wound, rather than by directly acting to kill bacteria. By causing mild local tissue injury, such caustic agents stimulated inflammation at the site of the wound. Such agents may also have directly stimulated the immune system.[16] The immune system called in to fight off the caustic substances would then be forearmed to attack any bacteria that tried to slip through the cracks of broken flesh.

In contrast to the Greeks, the Chinese used moldy tofu to treat inflammations and infections of the skin.[17] Similarly, the ancient Egyptians used moldy bread to treat skin lesions.[18] The use of moldy natural products by the Chinese and Egyptians as a cure-all for infections is relevant to modern antibiotics, since penicillin is produced by the bread mold *Penicillium*. Hence, ancient medicinal technologies thousands of years ago may well have been based on crude preparations of substances similar to antibiotics.

As civilization has advanced forward from the ancient era, physicians have continued to seek cure-alls for disease. But it was not until the twentieth century—three millennia after the Chinese, Egyptians, and Greeks had figured out how to use natural products as treatments—that the first signs of success in the long-standing quest for a cure-all began to come to fruition. It was the brilliant and legendary German scientist Dr. Paul Ehrlich—Nobel laureate for medicine in 1908—who first coined the term *chemotherapy*.[19] By chemotherapy, Ehrlich was referring to the long-sought desire to use specific chemicals to treat infections, cancer, or other diseases, without causing damage, or side effects, to the patient from the chemical treatment.

Ehrlich's own efforts in this regard led to the development of the first chemotherapy agent—the drug Salvarsan, a derivative of arsenic—to successfully treat syphilis. In 1910, when Ehrlich published his successful results using Salvarsan to cure syphilis,[20] worldwide demand for the drug skyrocketed. Unfortunately, Salvarsan did not quite measure up to Ehrlich's own definition for specific chemotherapy; while it was effective at killing the bacteria that caused syphilis, it was also quite toxic to the patient. So, the search for the magic bullet for infectious diseases continued, eventually leading to the discovery of antibiotics.

The word *antibiosis*—meaning "against life"—had been coined in 1889 by the French scientist Vuillemin, who grew mixtures of different species of bacteria and noted that one group of bacteria could exert an inhibitory effect on another group. Vuillemin wrote that antibiosis was the act of "one creature destroying the life of another in order to sustain its own."[21] The term had a varied and colorful history during subsequent decades, and no standard definition was widely accepted. In 1928, another pair of French scientists defined antibiosis as occurring when "one organism exerts an injurious effect on another." But how bacteria inflicted injury to other species of bacteria was totally misunderstood, leading to a general confusion of scientific terms used to describe the process.

It was in 1942, at the suggestion of an American scientist named Dr. Selman Waksman, that the word *antibiotic* came to be applied to chemicals that were produced by microbes and used to kill other microbes.[22] Waksman and his colleagues defined an antibiotic as "a chemical substance, of microbial origin, that . . . [inhibits] the growth or the metabolic activities of bacteria and other micro-organisms." Waksman would later win the Nobel Prize in medicine for his discovery in 1945 of the powerful antibiotic streptomycin, which was the first drug capable of treating tuberculosis.[23]

In 1928, fourteen years before Waksman coined the term *antibiotic*, a British scientist named Alexander Fleming made an important discovery. In what was, frankly, the result of sloppy laboratory work, he left a plate of bacterial culture in his laboratory sink, uncovered. He later came back and found that a mold, *Penicillium*, had started growing on the plate. Furthermore, bacteria had grown everywhere on the plate, except in the immediate vicinity of the mold, where no bacteria were growing. Fleming realized that a substance produced by the mold, which he called "penicillin," was killing off nearby bacteria. Penicillin was thus the first true antibiotic "discovered" by people.[24]

But Fleming himself never managed to isolate pure penicillin. He could only produce highly impure material that was not nearly as effective as his initial observation suggested it should have been. So, even though the impure material was potentially sufficient to cure infections in patients,[25] Fleming was unable to make significant headway in further isolating or characterizing penicillin. Frustrated by his lack of scientific progress, astoundingly, Fleming stopped working on penicillin and let this absolutely revolutionary discovery slip through his fingers so that it was almost lost to history.

It was not until 1940 that Howard Florey, Ernst Chain, and Norman Heatley, working at Oxford, revived the penicillin story. They were able to chemically isolate pure penicillin from cultures of the mold. They further demonstrated that the penicillin cured mice infected with what would have been lethal doses of the virulent bacteria *Streptococcus pneumoniae*, *Staphylococcus aureus*, and *Clostridium*.[26] By 1941, in collaboration with pharmaceutical companies that had essential expertise in fermenting cultures of microbes, the scientists were able to manufacture penicillin in a semipure form on a scale sufficient for use in people.[27] In the meantime, another revolutionary discovery had usurped penicillin's place in history as the first useful antibacterial magic bullet.

In 1927, Dr. Gerhard Domagk became the director of the research laboratory at the German company I. G. Farben, one of whose divisions later became the pharmaceutical giant Bayer.[28] Dr. Domagk was a pathologist by training, but his job at I. G. Farben was to supervise a laboratory that was exploring potential medicinal uses of chemical red dyes. He had the good fortune to have working for him two chemists, Fritz Mietzsch and Joseph Klarer, who had previously been involved with trying to find chemicals that could kill parasites. They thus had experience in trying to create poisons that were specific for pathogens (disease-producing organisms) but harmless to humans. Taking advantage of the chemists' phenomenal intuition, Domagk's team began chemically altering red dyes and testing them sequentially against bacterial infections. At some point in 1931—almost ten years before Florey and his team first cured bacterial infections with penicillin—Domagk and colleagues made an astonishing discovery.[29] They found that one of their dyes could cure otherwise lethal streptococcal infections in mice. They went on to chemically alter that compound, creating a sulfur-containing red dye called *prontosil rubrum*. By early 1932 they had determined that prontosil rubrum was an extremely effective cure for otherwise lethal bacterial infections both in mice and rabbits.[30] However, the dye could not kill bacteria in the test tube. Based in part on these observations, it was subsequently discovered that when prontosil was injected into mammals, their livers would metabolize the dye into the true, active drug, the sulfur-containing antibacterial drug, sulfanilamide.

Domagk and his colleagues quickly turned their basic science discovery into a treatment for patients. Although the results of their animal experiments were not published until 1935, news of the discovery leaked to the local media almost immediately after the initial animal studies were completed. As a result, immediate demand arose for access to the drug prontosil, and clinical testing actually began in late 1932, more than two years before the results of the animal studies were published. On May 17, 1933, a colleague of Domagk's reported using the new sulfa drug to cure a ten-month-old boy with what had been considered an inevitably fatal bloodstream infection caused by *S. aureus*.[31] This child was the first person in history ever to be cured of a lethal bacterial infection using a specific, antimicrobial magic bullet.

Subsequently, in 1935, Domagk's six-year-old daughter stumbled while going down the stairs and accidentally punctured her hand with a sewing needle. Within several days her hand swelled up, she spiked a high fever, and she became delirious. The infection spread through her lymph nodes, which swelled up in her arm to massive proportions. Her doctor tried to drain the pus by repeatedly sticking a sterile needle into the lymph nodes. The treatment did not work. Ultimately the doctor recommended amputation to save the child's life. In desperation, and with his daughter "near death," Domagk administered to her the very red dye his laboratory team had discovered and developed, prontosil rubrum. He wrote:

> At the beginning of the treatment on the morning of December 8 [1935] the axillary temperature was 39.3 degrees [102.7 degrees Fahrenheit]. The child was dizzy and without her taking notice of it I pressed the Prontosil tablets into her mouth, and in the afternoon I gave her 15cc, and in the evening 10cc of Prontosil Soluble rectally. Already the same evening the temperature fell to 38.7 degrees [101.7 degrees Fahrenheit]. . . . On December 12 the temperature had dropped to normal.[32]

Thus, in one of those great ironies that so frequently punctuate the marvelous annals of science, Hildegard Domagk, daughter of the creator of the first true antimicrobial agent, was one of the first people to have life and limb saved by the very same drug.[33] Later in 1935, the first person in the United States, a ten-year-old girl with a brain infection, received sulfanilamide.[34] Her life was initially saved by the drug, but several months later she relapsed and died.

Subsequently, the use of sulfa drugs exploded for a variety of life-threatening infections, such as bloodstream infections and pneumonia.[35] Physicians who used sulfa drugs to treat patients from 1936 through 1941 clearly understood the miraculous nature of the effects of these drugs.[36] It was as if the medical equivalent of the lightbulb had been invented, nothing short of a revolutionary change in the practice of medicine.

Physicians treating patients with sulfa drugs, and subsequently with penicillin, saw unprecedented rates of survival for infections that had previously killed most affected patients. No less a luminary than Winston

Churchill was saved from just such an infection by the new miracle of antimicrobial therapy. Churchill developed a life-threatening pneumonia during an extended diplomatic excursion in 1943.[37] At the time, Churchill was sixty-nine years old, and at that age, he had a 60 percent chance of dying from the pneumonia.[38] But his physicians treated him with a new sulfonamide drug and it cured his infection, allowing him to continue the planning for the D-Day invasion. With such an enormous global impact, it is not surprising that Nobel Prizes were awarded in 1939 to Domagk for his discovery of sulfa drugs (although the Nazis prevented him from accepting the award until ten years later) and in 1945 to Fleming, Florey, and Chain for their work on penicillin.

Since those initial heady days of antibiotic discovery and development, hundreds of different antibiotics of many different types have been discovered and developed. By 1948, more than five thousand sulfa compounds had been synthesized in the laboratory, of which at least seven were proven to be effective for treating infections in patients.[39] Also, as mentioned, Selman Waksman and his colleagues developed the revolutionary antibiotic streptomycin in 1945. The first antibiotic in the tetracycline class of drugs followed shortly thereafter, in 1948. Over subsequent years, new classes of antibiotics were derived from penicillins, such as cephalosporins, carbapenems, and monobactams. Entirely new classes of antibacterial drugs, completely unrelated to penicillins, sulfa drugs, or tetracyclines, were also discovered, such as macrolides, quinolones, aminoglycosides, lincosamides, streptogramins, rifamycins, glycopeptides, lipopeptides, and oxazolidinones. All told, more than 150 antibiotics of various different classes have been discovered and developed since 1935. Each new antibiotic has added to the overall ability of physicians to cure infections that previously would have killed patients.

Unfortunately, as each new antibiotic has become available, resistance to that antibiotic has inevitably followed in short order. And now, seventy years after the first cure of a patient with sulfa drugs, the rate of development of resistance to antibiotics has far exceeded the rate at which new antibiotics are being developed. We are at a crossroads in our struggle with infections. How we proceed over the coming decade is likely to decide the issue for the long term. The question before us is, do we want to try to keep up with antibiotic-resistant bacteria, and keep the medical miracle of antibiotics alive or not?

WHAT IS ANTIBIOTIC RESISTANCE?

A microbe is a living organism that is too small to be seen by the naked eye, and can only be seen through a microscope. There are many different types of microbes in the world, from bacteria to viruses to fungi to parasites. As we have discussed, an antibiotic is merely a selective poison that is designed to kill one or more of those microbes, but it remains relatively harmless to our bodies.

Every antibiotic has a specific "spectrum" of microbes that it is particularly good at killing. Whether or not an antibiotic is effective at killing a specific microbe depends on many different factors. For an antibiotic to kill a certain bacterium, for example, that bacterium must use the biochemical pathway that the antibiotic is designed to poison. Penicillin—the archetypal antibiotic—poisons the ability of bacteria to make their cell walls, which provide crucial structural support to bacterial cells. In the presence of penicillin, bacterial cell walls become porous, almost like a dike with no Dutch boy to plug the leaks. As the bacteria try to grow in the presence of their leaky cell walls, they become unstable and literally burst. But there are also bacteria that don't have cell walls, and other bacteria that make their cell walls by a different process that penicillin cannot block. Such bacteria cannot be killed with penicillin—they are inherently resistant to penicillin.

Another factor required for an antibiotic to kill a bacterium is that the antibiotic must be able to penetrate into the bacterium. Because some bacteria have thicker cell walls than others, or different types of cell membranes than others, different antibiotics may penetrate to varying degrees into those bacteria. Additionally, some antibiotics work only if there is oxygen present; other antibiotics may work better if there is no oxygen around, in what is called *anaerobic* environments.

Finally, bacteria can create self-defense molecules that literally destroy antibiotics, or block antibiotics from working correctly, or physically pump antibiotics out of the bacterial cell. These bacteria have developed specialized "resistance" molecules that are themselves encoded in the DNA sequence of special resistance genes, which allow the bacteria to defeat antibiotics. So there are many different types of antibiotic resistance, each of which is created as a result of one or more resistance genes contained within microbes.

How do we detect resistance when it is present in a bacterium? And how can we distinguish bacteria that are resistant to antibiotics from bacteria that are not? It's conceptually very simple. You try to grow the bacteria in the presence of an antibiotic. If it grows, it is resistant to that antibiotic. If it doesn't grow, it is not resistant, and it is instead called *susceptible* to the antibiotic. Of course, in the real world, testing for antibiotic resistance is a bit more complicated. Resistance is as much quantitative as it is qualitative. That is, resistance isn't so much a yes/no phenomenon as it is a continuous gradation effect. So, you actually have to grow the bacteria in many different concentrations of antibiotic to determine how much antibiotic is necessary to kill the bacterium. The more antibiotic necessary to kill the bacterium, the more resistant the bacterium is to that antibiotic.

Over the last seven decades, since sulfa drugs and penicillin first entered clinical use, hospital microbiology laboratories have become adept at growing bacteria in the presence of varying concentrations of antibiotics. These techniques have even been automated in high-tech machines that can run dozens of tests simultaneously on each bacterial sample. The lab technician takes a sample of bacteria from a patient's culture specimen and squirts the bacteria into a machine that simultaneously runs dozens of tests. By using these amazingly sophisticated machines, laboratories can determine precisely how much antibiotic must be present to inhibit a given bacterium's growth. This value—the Minimum Inhibitory Concentration (MIC) of an antibiotic for a bacterium—has become the standard method for determining if a bacterium is resistant to antibiotics.

Take, for example, *Streptococcus pneumoniae*, which historically has been known as "the pneumococcus." The pneumococcus has long been and continues to be the most common bacterial cause of pneumonia (lung infection) and meningitis (brain infection) in developed countries. This bacterium is usually transmitted outside of the hospital, and can even infect otherwise healthy, young people, causing dangerous lung, blood, or even brain disease. In the past, *S. pneumoniae* was extremely susceptible to penicillin. Specifically, less than one tenth of one millionth of a gram of penicillin dissolved in a cubic centimeter of water was sufficient to kill *S. pneumoniae* in the laboratory (i.e., the penicillin MIC was $\leq$ 0.1 microgram/milliliter).[40]

In 1964, two physician-scientists, Robert Austrian and Jerome Gold, published their experience using penicillin to treat bloodstream and lung infections caused by the pneumococcus.[41] Prior to the antibiotic era, the pneumococcus was such a deadly pathogen and so widely feared that the legendary physician Dr. William Osler had referred to pneumococcal pneumonia as "the captain of the men of death."[42] In another of those ironies that so frequently punctuate the annals of medical science, Osler himself ended up succumbing to that exact infection. Nevertheless, by twenty years after the availability of penicillin, Austrian and Gold were able to write that there was a general impression in the medical community that "pneumococcal disease no longer constitutes a serious medical problem."[43] Why would anyone think this? Austrian and Gold compared the death rates from pneumonia in the era of penicillin to the death rates in the preantibiotic era. They found that almost 90 percent of patients treated with penicillin survived pneumonia even when it had progressed to the point of having bacteria penetrate from the lung directly in the bloodstream (i.e., "bacteremic pneumonia"), versus a paltry 15 percent survival rate in the preantibiotic era. So impressed were they by the remarkable power of penicillin to destroy *S. pneumoniae* and cure patients of pneumonia that they concluded, "It is questionable that a more effective anti-pneumococcal drug than penicillin can be developed. It is bactericidal in very low concentrations, and no well-documented case of therapeutic failure as a result of pneumococcal resistance to penicillin has ever been reported."

Pretty heady praise, and clearly well deserved for penicillin at the time. Unfortunately, the times they are a-changin'. Over the last thirty years, *S. pneumoniae* strains have become more and more resistant to penicillin. In fact, in the year 2008, almost one-hundred-fold-higher penicillin concentrations may be required to kill *S. pneumoniae* than were required in 1964, when Austrian and Gold wrote up the results of their study.[44] This resistance of *S. pneumoniae* to penicillin has been caused by the bacteria evolving and adapting to chronic penicillin exposure. The bacteria have learned how to make their cell walls even though penicillin is present, thereby rendering the penicillin less effective.

This rising resistance can be a major problem when we try to use penicillin to treat infections caused by *S. pneumoniae*. Indeed, the ultimate definition of whether a bacterium is "resistant" or "susceptible" to an antibiotic is whether or not an infection caused by that bacterium can be cured by an antibiotic in patients.[45] However, numerous clinical studies have proven that laboratory testing in test tubes does predict the ability of antibiotics to cure infections in patients. That is, infections caused by bacteria resistant to antibiotics in test tubes have a much lower chance of being cured in patients than bacteria that are susceptible to antibiotics in test tubes.[46]

Society is in the midst of an emerging crisis of antibiotic resistance among microbial pathogens in the United States and throughout the world.[47] Epidemic antibiotic resistance has been described in many different types of microbes. For example, as mentioned above, penicillin resistance has been spreading rapidly among the bacterium *S. pneumoniae*. Remember that *S. pneumoniae* is the number-one cause of pneumonia and meningitis, and that these infections are usually transmitted outside of hospitals, frequently among otherwise healthy people. In some parts of the United States, an appalling 40 percent of *S. pneumoniae* strains are now resistant to penicillin and penicillin-derived antibiotics.[48] That compares with essentially 0 percent resistance to penicillin just two or three decades ago. More recent studies from Europe have found similar penicillin resistance rates among *S.*

pneumoniae, but even more concerning was that 25 percent of *S. pneumoniae* strains had become multi-drug resistant (that is, resistant to penicillin and at least two other classes of antibiotics).[49] Most recently, an outbreak of inner ear infections in otherwise healthy children in the United States was caused by a strain of *S. pneumoniae* that was "resistant to all FDA-approved antibiotics" that normally could be used to treat the infections.[50] The children had to have surgery to resolve their infections.

Another example of spreading drug resistance is the famous bacterium, *Staphylococcus aureus*, which causes what are known colloquially as "staph infections" of the skin. Thanks to recent press coverage, the term "MRSA"—methicillin-resistant *S. aureus*—has become fairly widely known even in nonmedical circles. I've already shared with you a story about one very unfortunate patient, Mrs. C., who had a life-threatening MRSA infection. But now it's time to delve more deeply into the frightening world of MRSA.

CHAPTER 3

methicillin-resistant *staphylococcus aureus*

deadly antibiotic-resistant bacteria escape the hospital

PENICILLIN: THE GREAT NEW HOPE AGAINST INFECTIONS

The bacterium *S. aureus* was once totally susceptible to penicillin. Before 1941, virtually all *S. aureus* could have been killed by penicillin. In fact, the first patient ever to receive pure penicillin treatment was a forty-three-year-old British police officer named Albert Alexander, who had an *S. aureus* infection. On October 12, 1940, poor Officer Alexander was admitted to an infirmary at Oxford, England, with a raging staph infection on his face.[1] The infection had started as a mild sore "at the corner of the mouth a month earlier." It had inexorably spread to cover his face, scalp, and both eyes, and also spread down his neck to his right upper arm. Nothing, it seemed, could stop the infection, which progressed despite sulfa treatment, and despite multiple surgeries to drain abscesses and pus. The infection rotted Officer Alexander's left eye so profoundly that his doctors had no choice but to surgically remove the dead eye. Shortly thereafter Officer Alexander's right eye required surgery to drain pus. The following day he began coughing up pus from his lungs, as the infection continued its inexorable spread.

Finally, in what was a last-ditch, desperate attempt to save his life,

Officer Alexander's doctors gave him a low dose (200 mg) of experimental penicillin through the vein. Subsequently several more doses were given. The following day, the doctors wrote in Officer Alexander's chart, "improvement after total of 800 mg penicillin in 24 hours. Cessation of scalp discharge, diminution of right-eye suppuration and conjunctivitis. Arm discharge seemed less." Shortly thereafter Officer Alexander's fever broke, and he maintained a normal temperature. By day five, the physician-scientists noted, "Total [penicillin] administered, 4.4 [grams] in 5 days. Patient felt much improved; no fever; appetite much better; resolution of infections in face, scalp and right orbit." Officer Alexander was heading toward an utterly *miraculous* recovery!

Unfortunately, by day five, the small supply of penicillin that had been produced in the laboratory had been exhausted. In fact, the supply had been exhausted by day three, but, determined to continue giving the drug to Mr. Alexander, the dedicated physician-scientists started the "P patrol," repurifying the penicillin from the patient's urine so they could administer it back into his veins. This urinary recycling of penicillin bought the doctors another two days of therapy before the amount of penicillin recovered in the urine was insufficient to continue treatment. Then the drug ran out.

Officer Alexander did well for ten days off therapy. Then, suddenly, his pneumonia came roaring back and took a deadly grip of him. Lacking any more penicillin, Officer Alexander died several weeks later. At autopsy he was found to have staph infection raging throughout virtually every part of his body.

Fortunately, subsequent experiences with penicillin were more successful.[2] In February of 1941, penicillin caused a dramatic cure in a fifteen-year-old British boy who had developed a severe streptococcal infection at the site of an otherwise simple surgery on his leg. Before the penicillin treatment, the young man "looked ill, pale and wasted." He was facing a real possibility of amputation to save his life. But then the doctors noted that "Penicillin therapy was followed by a great improvement in the patient's general condition." In the preantibiotic era, surgery even of the simplest type was commonly deadly due to resultant infections. The availability of antibiotics revolutionized the field of surgery, and early hints of the promise of antibiotics to make an enormous difference were evident in cases such as the one of this young man.

On May 3, 1941, another man with a staph infection on his back received penicillin. The infection melted away. There are several other cases of treatment success described in this first publication on the use of systemic penicillin treatment of human patients. Again, in reading the doctors' descriptions of these treatment responses, one gets the overwhelming impression of awe on the part of the observers. These doctors knew they had harnessed a power heretofore unseen in the millennia-long annals of medicine, as if drawing forth and bottling manna from the heavens.

Then, in the following year, in 1942, a thirty-three-year-old American woman, Mrs. Anne Sheafe Miller, was delirious and on the brink of death due to a severe streptococcal infection.[3] She was being cared for at Yale–New Haven Hospital in Connecticut. She had failed to respond to sulfa antibiotics, and had been continuously running fevers in excess of 103 degrees Fahrenheit for more than four weeks. It just so happened that Dr. John Burton, an old schoolmate and close friend of the developer of penicillin, Dr. Howard Florey, was also hospitalized at Yale–New Haven with an unrelated illness. Mrs. Miller's physician persuaded Dr. Burton to contact Dr. Florey and request that penicillin be made available to try on Mrs. Miller.

According to Dr. Charles Grossman, an eyewitness who was directly involved with the care of Mrs. Miller, a small amount of penicillin was delivered from the pharmaceutical company, Merck, to the hospital on Saturday, March 14, 1942.[4] Dr. Grossman and his colleagues dissolved the penicillin powder in a saline solution and injected it into Mrs. Miller's IV. Miraculously, and literally overnight, Mrs. Miller's fever dropped and her delirium resolved. She went on to a complete cure and recovery. So powerful was the effect, and so amazed were Mrs. Miller's physicians, that one of them, a senior consultant named Dr. Wilder Tileston, was accidentally overheard muttering to himself, "Black magic," while he was reviewing her chart!

Anne Sheafe Miller was thus the first person in the United States to be cured of an infection by penicillin, and arguably was the first person ever to have her life saved by the miracle drug. How big of an impact did penicillin make for her? Well, having been cured by the antibiotic, Mrs. Miller went on to survive to the age of ninety, and raise children and grandchildren. She died only quite recently, in 1999.[5] Penicillin bought her more than a half-century prolongation of her life. What other medical intervention, in the entire catalog of the pharmacopoeia, can lay claim to that power?

The first published treatments of skin infections with penicillin in the United States came in 1942 and 1943, by Dr. Herrell from the Mayo Clinic.[6] How spectacular were the successes? Try to imagine that your four-year-old daughter, who had previously been in excellent health, suddenly develops a skin infection on her face and neck. The infection spreads relentlessly, and within four days your daughter spikes a fever of 104 degrees Fahrenheit, and she cannot sleep because her face and neck are so swollen, she can hardly swallow her own saliva. She begins gasping for breath. You rush her to the hospital, where doctors drain some pus from her face and grow the dreaded *S. aureus* bacterium. The doctors find evidence that the infection has spread both to the lung and into the blood. You are out of your mind with fear, and your doctors have nothing reassuring to say; they make it clear that they expect your daughter to die within two or three days, and that there is nothing they can do about it.

This is no hypothetical case. The four year-old-girl I just described was actually brought to the Mayo Clinic in 1942. Dr. Herrell wrote that when she arrived at the hospital, she was "moribund," meaning on the verge of death. She did not respond to surgical drainage of pus. Fortunately, Dr. Herrell was at the cutting edge of medicine and had access to a small quantity of penicillin. He administered a tiny dose of penicillin into the girl's vein (20,000 units on day 1 and 30,000 units on day 2—compare that to a dose of one to two *million* units every four hours, which is how the drug would be dosed in the twenty-first century!). Despite the

Figure 3.1. A Moribund Four-Year-Old Girl with a Staphylococcal Infection, Who Was Cured with Low Doses of Penicillin (reprinted with permission from W. E. Herrell, "Further observations on the clinical use of penicillin," *Proceedings of the Staff Meetings of the Mayo Clinic* 18 [1943]: 71, Dowden Health Media). The panels on the top show the girl shortly after presentation to the hospital. The panels on the bottom show her after a two-week course of penicillin therapy.

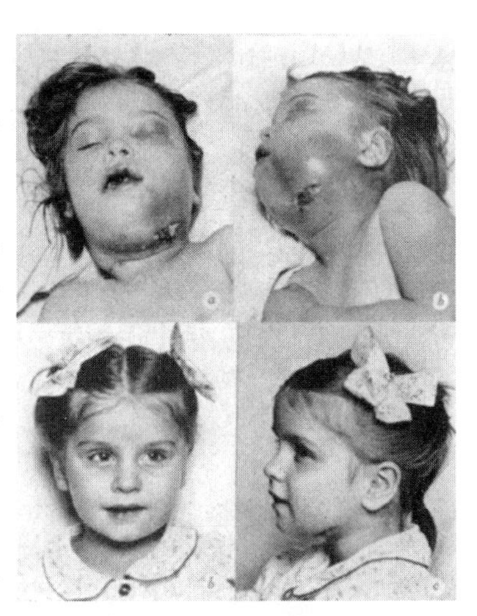

laughably low doses of penicillin, within thirty-six hours, the girl's fever began to dissipate, her pneumonia resolved, and her blood cultures cleared of bacteria. By four days after initiation of treatment, the swelling of her face and neck were so much better that she was able to swallow normally. By the end of a two-week course of penicillin therapy, this little girl, whom Dr. Herrell had considered to have an "almost universally fatal" infection, was completely cured and appeared totally normal (fig. 3.1). Black magic indeed!

According to a personal communication from Dr. James Steckelberg, chief of infectious diseases at the Mayo Clinic, the little girl whose pictures are shown in figure 3.1 is now a healthy woman who is alive and well. Penicillin has thus far given her a seven-decade lease on life, and counting! It is through countless absolutely miraculous stories such as these that the true power of antibiotic therapy came to be realized. And, unlike in the modern era, the physicians and patients of the 1940s and 1950s *appreciated* that power. They knew how bad it had been before penicillin, and they understood what the availability of antibiotics meant for patients and for society.

For example, in 1950, a grateful mother wrote a letter to the editor of *Lancet*, one of the most prestigious medical journals in the world. I do not believe that many laypeople are prone to writing letters to medical journals, nor do I believe many prestigious medical journals print such letters. But this letter was striking, and it is not surprising that it got published:

> As a parent who has just seen, once again, the effect of penicillin on a sick child, may I put on record the immense gratitude that one feels to all those known and unknown chemists who have made this miracle possible? And not only to these but to the whole medical profession who, directly and indirectly, do so much in guarding our happiness including myself.
>
> It seems such a long way from the nursery to the medical school and the laboratory that gratitude tends to become inarticulate; but nevertheless it is very deeply felt by a great number of people.[7]

THE MICROBES STRIKE BACK!

As the manufacturing capacity of penicillin improved, subsequent use of penicillin indeed proved revolutionarily successful, and the antibiotic era blossomed on the backs of sulfa drugs and penicillin. Unfortunately, the era of antibiotic resistance was born shortly thereafter, as the first infection caused by penicillin-resistant *S. aureus* was observed within a year or so of the first clinical use of penicillin.[8] Astoundingly, it took only ten years for penicillin resistance among *S. aureus* to spread throughout the entire United States (fig. 3.2) and the world.[9] This penicillin-resistance problem started in hospitals, where the highest density of penicillin use was occurring. But within twenty years, penicillin resistance among *S. aureus* had escaped the hospital and began spreading like wildfire.

It took Western medicine 2,500 years (from the time of Hippocrates) to discover antibiotics and gain the upper hand on bacterial infections. Then, almost immediately, microbes made a quantum leap ahead of medicine by becoming resistant to our antibiotics. But, within a decade, pharmaceutical companies struck back once again on behalf of humanity. In the late 1950s, pharmaceutical companies discovered new versions of penicillin derivatives that could overcome *S. aureus*'s resistance mecha-

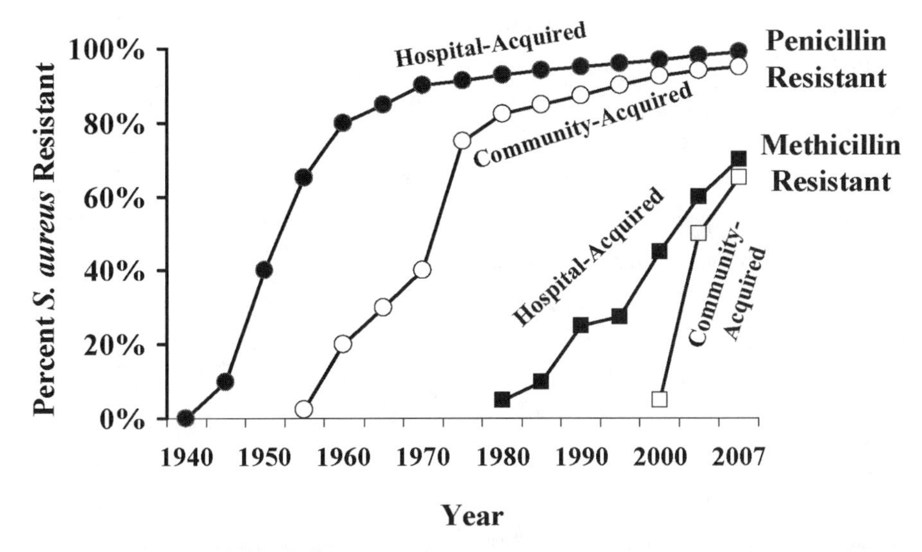

Figure 3.2. The Spread of Antibiotic-Resistant *S. aureus* Infections in the United States (adapted from H. F. Chambers, "The changing epidemiology of *Staphylococcus aureus*?" *Emerging Infectious Diseases* 7 [2001]: 178–82, with permission). Shown on the y axis is the percentage of *S. aureus* bacterial strains that were resistant to penicillin or the newer drug, methicillin, plotted year by year in the United States (x axis). Separate curves are shown for bacteria that caused infections in the hospital ("Hospital-Acquired") and in healthy people in the community ("Community-Acquired"). The graph shows that resistance to penicillin and methicillin started in hospitals and only later began to spread outside hospitals.

nisms. A family of these penicillin derivatives, including methicillin, oxacillin, nafcillin, and so on, entered widespread clinical use beginning in 1961. These new penicillin derivatives were extraordinarily effective at killing penicillin-resistant *S. aureus* and became the mainstay of treatment of staph infections for the next thirty years. But resistance to these new miracle drugs was also seen within a year of their widespread use (fig. 3.2). As for the prior emergence of penicillin resistance, this resistance to new penicillin derivatives first appeared among hospitalized patients. From the 1970s through the 1990s, so-called methicillin-resistant *S. aureus* (MRSA) exploded into hospitals. While these MRSA infections wrought havoc in hospitals throughout the developed world, it was very fortunate that such drug-resistant infections were virtually never seen in infections acquired outside the hospital.

All of that began to change in the late 1990s, when MRSA suddenly leapt out of hospitals and began causing community-acquired infections across the United States, Europe, Australia, and Asia (fig. 3.2). Over the last decade, a pandemic of hypervirulent, antibiotic-resistant MRSA infections has been raging across the globe.[10] At my hospital, Harbor-UCLA Medical Center, we virtually never saw an MRSA infection from the community prior to 1999. All of our MRSA infections were hospital acquired. Then, in 2000, during my residency in internal medicine, the floodgates began to open, and patients like Mrs. C. inundated our hospital. By the end of my residency, in 2002, we were seeing several cases of MRSA infections from the community *every day*—that is, a thousand cases a year or more, just at my hospital. We went from no infections per year to several infections per day in a span of less than four years. And that rate of infection continues to this day.

It is ironic that there has been significant press coverage of the MRSA threat recently, as if the problem is new, or newly discovered. In 2007, the press jumped on a research article and accompanying editorial published in a leading medical journal, which estimated that ninety thousand cases of invasive MRSA occur per year in the United States, resulting in approximately twenty thousand deaths, which is more deaths than are caused by HIV in the United States every year.[11] The launching of a firestorm of news coverage based on these studies was interesting, since this concept was not at all new to infectious-diseases specialists. Nor was the scale of the problem new. The new research updated the numbers, but the general scope of the problem has been known for years. In fact, my response to the press coverage was, "No duh!"

Furthermore, the new research really addressed only the most severe forms of infections caused by MRSA. Several years earlier, a leading *S. aureus* expert, Dr. Henry "Chip" Chambers from the University of California at San Francisco, estimated that if one took into account all infections caused by *S. aureus*, the true number is probably 1.8 million infections every year just in the United States![12] Recent data from his group validate that number.[13] Of those, an astonishing one million infections per year are now caused by MRSA. While the majority of these infections are skin abscesses, requiring a minor incision to drain the pus followed possibly by a brief course of antibiotics, perhaps 5–10 percent of the infections do become more invasive, including infections of the blood and other internal organs.

Most of these MRSA infections occur outside of hospitals, in otherwise healthy people in communities. In many areas in the United States, MRSA is now endemic, meaning that drug-resistant *S. aureus* has replaced drug-susceptible *S. aureus* as the leading cause of staph infections in either the community or hospital settings.

Which brings us to Mr. D.

Mr. D. was a pleasant young man of about thirty who came to the hospital with fever, complaining of pain and swelling in his side. I met him in the emergency room, having been called to see him by his admitting internal-medicine resident, who wanted a consult from an infection specialist. I first saw Mr. D. at about five o'clock in the afternoon. He was smiling and appeared comfortable as I introduced myself to him. I asked him if he had cut or otherwise injured himself prior to the infection occurring.

"No, nothing like that," Mr. D. said. "I've been fine. I really don't have any idea how this happened."

I asked him to lean forward so I could examine the infection. On his right side and back I saw an area of inflammation, perhaps four inches in diameter. It appeared to be a "cellulitis" (or infection of the skin and underlying soft tissue). But what was specifically concerning about Mr. D.'s cellulitis was its color. Most cellulitis appears red. In contrast, Mr. D.'s cellulitis was dusky purple. Dusky purple makes an infectious diseases doctor nervous, because it can be seen with deeper types of infections, such as necrotizing fasciitis (the dreaded "flesh-eating bacteria" of tabloid fame). But Mr. D. did not appear to be sick enough to have necrotizing fasciitis, and his blood tests didn't look that bad. Plus, Mr. D. was

African American, and the redness of cellulitis can be less obvious in darkly pigmented individuals. So I decided to treat Mr. D. for cellulitis, using an antibiotic that would deal with the most concerning bacterium that might cause a skin infection. At the time, our hospital was already awash in community cases of MRSA skin infections, so I naturally started Mr. D. on vancomycin, the standard antibiotic treatment for serious MRSA infections.

The next morning I came back to see Mr. D., who had been admitted upstairs overnight. I saw him at about eight in the morning. In the intervening sixteen hours, Mr. D. had begun to feel very poorly. As I walked into the room, I could tell from the doorway that he was in distress. He was grimacing and breathing heavily in his bed.

"How're you feeling?" I asked.

"It's much worse, Doc."

"More pain?"

"Yeah. Also I had a high fever last night. And I can't seem to catch my breath."

I glanced down at the vital sign sheet at the foot of Mr. D.'s bed. My eyes widened as I saw the temperature reading from that morning, 105 degrees Fahrenheit. I glanced at the electronic monitor over his bed, which showed that Mr. D.'s heart rate had shot up overnight. He was also breathing much more quickly than before, which was a sure sign that something was wrong. With a bad feeling in my gut, I stood there watching Mr. D.'s automatic blood pressure cuff cycle. I waited for the cuff to inflate and then deflate. It beeped. Ninety over forty flashed on the monitor. Uh oh, I thought. Mr. D.'s blood pressure was falling.

"Let me have a look," I said. "Can you roll on to your side?"

Mr. D. gritted his teeth as he turned, the pain intense. I uncovered his back, moving the hospital gown out of the way. What I saw stunned me, and I fought to keep my jaw from dropping.

Overnight, Mr. D.'s infection had rampaged up his back, across his flank, and down his thigh. The infection had spread from four inches to perhaps three feet in diameter in sixteen hours. Cellulitis doesn't do that. Necrotizing fasciitis—the dreaded "flesh-eating disease"—does.

I bit off a swear word as I covered Mr. D. with his gown. "Okay, you can lean back."

"So, how's it look?" he asked, trying to catch his breath.

"Well, not so good." His eyes widened a little. "Listen, I'm going to call the surgery team to come here and have a look at you."

"Surgery? I need surgery?" He was panicking.

"Yes, I think so. Don't worry, our surgeons are very experienced with these types of infections. Just hang in there and sit tight for a minute, okay? I'll be right back."

I grabbed Mr. D.'s chart and rushed to the phone. As I paged the on-call trauma surgeon, I flipped open the chart. I was certain that Mr. D. had necrotizing fasciitis. So this was nothing like Mrs. C.'s case had been. Mrs. C. had had a spinal infection, which was commonly caused by MRSA. In contrast, Mr. D. had necrotizing fasciitis, which was known *not* to be caused by MRSA. Since I had been treating Mr. D. with vancomycin, an antibiotic specific for MRSA, and MRSA was not a known cause of necrotizing fasciitis, I was thinking I had given Mr. D. the wrong antibiotic. Necrotizing fasciitis has a high mortality even with appropriate treatment, and with the wrong antibiotic, it would almost certainly be fatal. I was feeling terrible that I might have given Mr. D. the wrong antibiotic. I wrote orders for several more antibiotics and called the nurse over.

"Fax these antibiotics orders down to pharmacy, STAT, please. And tell the pharmacy to send a runner up here with them as soon as they are ready. As soon as they get here, please hang the IV. We've got to get these antibiotics into him right away."

The nurse hurried to the fax machine.

When the on-call surgeon called back, I explained the situation. The trauma team immediately came to evaluate Mr. D. It took them less than a minute to examine him. "Looks like nec fasc," the chief resident agreed. "We're prepping him for the OR. We'll take him right now."

They whisked Mr. D. down to surgery. In the operating room, Mr. D. was indeed found to have necrotizing fasciitis. His entire back, side, and leg were filleted open like a side of beef in an attempt to remove as much of the infection as possible. The surgeons found giant pockets of pus that had eroded into Mr. D.'s muscles, and rotting flesh described as "dishwater necrosis" because the liquefying tissue has that classic, brackish dishwater appearance. But at least Mr. D. was alive.

The next day the microbiology laboratory posted the results of Mr. D.'s cultures, taken by the surgeons in the OR from his wound. I was once again shocked as I read the computer screen: MRSA. Necrotizing fasciitis from MRSA? That just wasn't supposed to happen.

I vividly remember standing in the surgical intensive care unit examining Mr. D. the following morning. Mr. D. was unconscious, on a mechanical ventilator to breathe for him. He was in florid sepsis, meaning that his blood pressure had plummeted due to the severity of infection. He was drenched in sweat, disheveled, and was swollen like a beach ball from all the saline that had been pumped into his body to keep his blood pressure up. He was barely recognizable. His butchered flank, back, and leg were covered in layers of gauze, which were stained brown and red from the brackish discharge slowly oozing into them from his massive wounds.

The trauma service was rounding on its post-op patients. They stopped outside the room as I was examining Mr. D. The door was open and I could hear them just a few feet away. I'll never forget the scene. There was a UCLA medical student rotating through the trauma surgery service. As the on-call resident read off the results of Mr. D.'s wound culture, the medical student said, "But I thought *Staph aureus* doesn't cause necrotizing fasciitis!"

The medical student was correct. At least, that's what all the text books said.[14] Necrotizing fasciitis is supposed to be caused by other types of bacteria, but not *S. aureus*, and certainly not MRSA (i.e., drug-resistant *S. aureus*). When I had first seen Mr. D.'s expanding infection, the morning after his admission, I had been worried that I had given him the wrong antibiotics, since I had been targeting MRSA rather than necrotizing fasciitis. It turns out I had given Mr. D. the right initial antibiotic after all! In Mr. D. I was treating *both* necrotizing fasciitis and MRSA.

There were virtually no reports in the world's literature of *S. aureus* as a sole cause of necrotizing fasciitis before 2005. Yes, *S. aureus* on occasion had been cultured from cases of necrotizing fasciitis, but only when other bacteria that were known causes of necrotizing fasciitis had also been cultured. So, it was simply axiomatic: *S. aureus* doesn't cause necrotizing fasciitis. Not only was *S. aureus* not supposed to cause necrotizing fasciitis, it was literally unheard of for MRSA to do so.

But bacteria don't read textbooks. This strain of MRSA had caused necrotizing fasciitis. It was also an unusual form of necrotizing fasciitis. Most patients who have necrotizing fasciitis appear to be deathly ill when they come to the hospital. Mr. D. had not initially appeared that ill. His infection was ripping through his skin and deeper tissue, but he had not appeared systemically ill, at least not until the next morning. In this way, Mr. D. did have something in common with Mrs. C. from chapter 2. These cases illustrate an important difficulty that doctors face in treating invasive MRSA infections. For some reason, MRSA seems to have the ability to cause very severe, deeply invasive infections but at the same time the patient may not initially appear very sick. This feature can make it very difficult to diagnose invasive MRSA infections or appreciate their extent and severity. We will see this feature present itself repeatedly as we discuss MRSA.

Mr. D. was on the verge of death after his initial surgery, but fortunately, ultimately he survived. Nevertheless, he spent weeks in the hospital and had to undergo multiple intensive surgeries to keep his infection under control. He was left with a gaping wound in his thigh and back that slowly closed up over months of healing.

After being confronted with Mr. D.'s highly unusual case of MRSA necrotizing fasciitis, my colleagues and I began an investigation. We reviewed all of our necrotizing fasciitis cases over an eighteen-month period. What we found was shocking. At our hospital in Los Angeles, fully one-third of all cases of necrotizing fasciitis were now being caused by MRSA.[15] These cases were typically being discovered in the operating room. Often the diagnosis had not been suspected until the surgeons opened the skin layer and followed a trail of pus and dying, rotting flesh deeper and deeper and deeper, all the way through the last skin layer, through the so-called fascial plane of tissue, and into necrotic muscle.

These cases of necrotizing fasciitis caused by MRSA were the ultimate example of the convergence of antibiotic resistance with increased virulence.[16] Not only was this strain of *S. aureus* hard to treat because it was resistant to first-line antibiotics, but it had also acquired the ability to cause completely new manifestations of disease. Prior to 2005, the textbooks specifically recommended not treating for MRSA when treating necrotizing fasciitis. The textbooks have now been rewritten.

People who previously have been completely healthy can come to the hospital with rip-roaring MRSA infections in the blood, on heart valves, in bones, in the spinal cord resulting in paralysis, and in the brain. The infection is now established in the community. It is passed by contact with people who carry the bacterium on their skin. A handshake, or similar contact, may be all that is required to transmit it, and it is impossible to know who is carrying it and who is not. Even worse, it can be passed by *fomites*—that is, inanimate objects touched by carriers of the bacterium.[17] A doorknob, a telephone, a car door, silverware, a pet's nose . . . any object has the potential to serve as a fomite for MRSA transmission. Why do some people carry the bacteria and not seem to get infected, while others get infected right away? No one knows. How do we stop transmission? No one knows. The reality is, anyone can acquire MRSA at any time. No one is safe.

Another factor that makes MRSA infections so scary—aside from the facts that they can be passed by mere contact and that they can present initially with subtle symptoms that do not reflect the true severity of the infection—is that healthy, young people in the prime of their lives have been devastated by these infections. Some of the earliest community MRSA outbreaks occurred among military personnel, including shipwide outbreaks among the crews of US naval vessels.[18] MRSA is also a frequent cause of infection in recruits during US Army basic training[19] and among military personnel at both army and air force bases.[20] Similarly, young athletes, particularly those playing contact sports, are prone to MRSA infections. Outbreaks have occurred in major college football programs,[21] and highly publicized outbreaks occurred among the professional St. Louis Rams team[22] and among suburban high-school athletes participating in contact sports, including football, fencing, wrestling, and so on.[23] What these outbreaks have in common is close-quarters contact in locker room or barracks settings that may not be conducive to maximal sanitation, mixed with frequent minor nicks, cuts, and bruises that may allow the *S. aureus* bacteria to slip through cracks in the skin to cause infection. And because these infections occur outside of hospitals, doctors often have a hard time realizing that the infections are being caused by a drug-resistant form of *S. aureus*.[24]

The highly publicized outbreaks of MRSA infections are dwarfed by the enormous number of cases that occur every day across the United States and throughout the world. Overall, healthy children, adolescents, and teenagers have been particularly heavily hit by MRSA infections,[25] and these cases had gone unheralded until very recently. As mentioned, MRSA infections have run wild through high school and college sports teams[26] and, frankly, are occurring at alarming rates even among the general population,[27] often with devastating effects.

In late 2007, a flurry of news media reports documented outbreaks of MRSA infections in high schools throughout the country. For example, in October of 2007 in Virginia, newspapers and television news programs reported that a seventeen-year-old high school senior named Ashton Bonds was killed by an MRSA infection.[28] Ashton's infection started as MRSA infections so often do, with a simple skin bump. Several days later he developed pain in his side. Within three days, the infection had spread via the blood to multiple internal organs. The previously healthy young man was dead within two weeks. As a result of the uproar over his death, as well as infections in other high schools, the state closed twenty-one high schools and spent a week thoroughly cleansing and sanitizing common areas before reopening.[29] Similarly, in Pennsylvania thirteen members of one high school's football team[30] and five football players at another high school were diagnosed with MRSA infections.[31] Elsewhere in Pennsylvania, cases of MRSA were reported in high schools in seventeen different school districts.[32] In North Carolina, six players on a high school football team were diagnosed with MRSA.[33] Amazingly, each of these news stories was written within a two-month period toward the end of 2007.

The Infectious Diseases Society of America, an international society of more than 8,600 infectious-diseases researchers, private practitioners, and leading academic physician-scientists, has been gathering examples of these tragic cases on its Web site (http://www.idsociety.org/STAARAct.htm) to try to help publicize the problem by putting a human face on the disease.[34] On the Web site, you can read about Rebecca Lohsen, a seventeen-year-old, completely healthy, honor roll student-athlete, who out of the blue developed MRSA pneumonia and died despite maximal medical therapy. You can read Mrs. Lohsen's description of her daughter's fight to stay alive. Mrs. Lohsen tells us that Rebecca's doctors believed

> they had her on the right antibiotic from the start and this should be no more difficult than any other pneumonia of this kind. And I had confidence that in this day of modern medicine they could fix almost anything, couldn't they? But her CAT scans continued to worsen; someone said her lungs looked like Swiss cheese. Then . . . they needed to insert a tube in her airway because they could no longer keep her oxygen levels up.

Mrs. Lohsen's story continues with every parent's worst nightmare,

> that awful night when I awoke in [Rebecca's] room (I never left her side) to a flurry of activity that was all too familiar to me from my days as an ICU nurse: Rebecca was "coding"—going into cardiac arrest. I don't know how I ever got out of that room. I only remember standing in the hall, hysterical—how could this be happening!

Tragically, Rebecca lost her struggle with MRSA. She died despite the best efforts of her treating physicians. And Rebecca's is not the only disastrous story you can find on the IDSA's Web site. Mrs. Amber Don writes about how the same disease, MRSA pneumonia, killed her completely healthy thirteen-year-old son, Carlos. Mrs. Don writes that "Carlos was the picture of health. He was very involved with sports and anything he tried came naturally to him. . . . Football was Carlos' best sport. His coach compared him to a gazelle. . . . Carlos could outrun anyone." And then Carlos went on a school camping trip. He left feeling fine. He came back four days later with rip-roaring pneumonia that developed completely out of the blue. Only a few days later he died in an intensive care unit, despite maximal medical therapy.

Read Mrs. Don's words:

> Carlos left us . . . while his father and I held his hand and told him over and over again how much we loved him. Our lives will never be the same. I miss him every single minute of every single day. Pictures and memories are all I have left of him and you can't give those hugs or tuck those in bed at night. The day I picked up his urn from the mortuary I also picked up my daughters from school. While waiting in my car for the girls I sat and watched my son's friends laughing and playing around outside the school. While they were doing what normal twelve-year-olds do, my son's remains sat in a box in the back seat of my car. He should have been out there laughing and playing with them.

You can imagine how totally devastating it is for a family to lose a child in this way. You can imagine how Mrs. Theresa Drew must have felt losing her twenty-one-year-old son, Ricky Lannetti, a star football player at Lycoming College. "Ricky," his mom writes, "was as strong as an ox and he ran like a deer. Best shape he would ever be in." Then he developed "flu-like" symptoms the week before a playoff game. But it wasn't the flu, and Ricky became very, very ill. He was hospitalized and put into intensive care. That night he died, despite receiving five antibiotics. His autopsy revealed that he had developed systemic infection caused by MRSA, which had entered his body through a simple skin infection. There was no rhyme or reason for it. It just came and killed him.

There are still more stories on the Web site. Brandon Noble, a professional football player for the Washington Redskins, writes of how he lost his career to a terrible struggle with MRSA infection of his knee. Mrs. Everly Macario writes of how her fifteen-month-old baby, Simon, died of MRSA pneumonia. Dee Dee Wallace writes of her near-death experience with MRSA necrotizing fasciitis, just like that of Mr. D. You can also read about young Bryce Smith, who developed MRSA pneumonia at fourteen months, and then developed hospital-acquired infections on top of the pneumonia. Bryce survived, thankfully, but only after months of harrowing struggle and a hospital bill that topped the one-million-dollar mark.

All of these patients developed their initial infections in the community, not in the hospital. All were in perfect health before they got infected.

As I said, everyone is at risk.

Everyone is at risk. . . .

In March of 2008, an otherwise healthy thirty-four-year-old male developed an abscess on his face that came on suddenly. The patient had gone to sleep the prior evening feeling fine. He had awoken at about three o'clock in the morning feeling a pressure sensation in his face. He looked in the mirror and was very alarmed to note a tender swelling on his jaw. Being familiar with the epidemic of MRSA infections in the city where he lived, the patient sought immediate medical attention, notifying his physician at an ungodly hour of the morning.

Okay, I admit it. The patient, and the doctor he notified, is me. Remember when I said that everyone is at risk? Well, I do mean *everyone*. This infection happened to me as I was in the late stages of writing this book. It hit me completely out of the blue. I have no idea why the infection occurred when it did. I've been around and among MRSA, and patients infected with MRSA, for years. I've performed physical examinations on innumerable patients with MRSA infections. I've personally cultured some of these patients. Furthermore, I studied MRSA in the laboratory as part of my vaccine experiments for several years, handling the bacteria on a regular basis. And yet I was never infected before. Why now? What changed? I never noticed a specific cut, from shaving or otherwise, on my face. So, really, I have no idea why it happened when it did. It was just my time.

I was not what you would call thrilled by this development. You see, I know how bad MRSA can get. I've just finished telling you all about it. Yes, I know that most of these infections respond briskly to drainage and antibiotics. But I also know that 5–10 percent of cases become more invasive. And this swelling had come on so quickly that I was very concerned about its rate of spread. Not to mention, it was on my face! I was taking no chances.

At three in the morning I drove to work, drained several milliliters of pus from the center of the abscess using a syringe needle, and phoned a 24-hour pharmacy to immediately prescribe myself an antibiotic. I cultured the organism from the pus I had drawn, and found that it was, indeed, MRSA. For the next three days, my face looked like I had been beaten in a bare-knuckle boxing match, or perhaps as if I had a really bad dental abscess on that side. The next day I had to drain more pus that had accumulated, but it was a smaller amount. By day two the infection had begun to respond to the combination of drainage and antibiotic. It took more than a week to completely resolve.

Do you know what I was thinking about that first morning, when I awoke, turned the light on, and looked in the mirror? I was thinking about Albert Alexander, the police officer who died from a facial staph infection in 1940, on the cusp of the antibiotic era. His infection had started in a similar location to mine. Had I developed this infection in 1940, Officer Alexander's fate could have been mine. Had Officer Alexander developed his infection just a year later, he almost certainly would have survived. I was also thinking about the four-year-old girl at the Mayo Clinic whose life was saved by penicillin from an otherwise lethal facial infection caused by *S. aureus*. I was also thinking about Mr. D., who developed necrotizing fasciitis from MRSA. I was thinking about all the other patient stories I've shared with you in this chapter. And, finally, I was thinking, "I am extremely thankful that there are still antibiotics around that are effective at treating this infection."

To paraphrase the old adage, I'm not only a prescriber of antibiotics, I'm a satisfied customer as well. I only hope that purveyors and customers alike will continue to be able to access effective antibiotics in the coming years.

If there is a bright side to the MRSA story, it is that drug companies saw the potential of MRSA to cause health problems very early during the hospital epidemic. That means that drug companies had a twenty-year head start on the problem that ended up spilling into the communities. By the mid-1980s, pharmaceutical companies understood that currently available antibiotics were inadequate for treating hospital-acquired MRSA infections. So they began to try to address the problem by developing new drugs for MRSA. After investing billions of dollars in such development, in the last decade pharmaceutical companies have brought to market several new antibiotics with which to treat *S. aureus* infections. There are also some older antibiotics that were previously rarely used to treat *S. aureus* infections that we have "rediscovered" and that can be used against these new strains of MRSA. Therefore, even as MRSA exploded into communities, we have had some drugs with which to treat it. But, do not be fooled. MRSA will become widely resistant to these antibiotics as well. It is inevitable. And it has already started. Resistance to the newest anti–*S. aureus* drugs has already been described, with the first cases occurring within a year of the new drugs' release.[35]

CHAPTER 4

beyond mrsa

infections resistant to virtually all antibiotics

Mr. A. is the index case. He's the source. He's the one that brought the drug-resistant strain of the bacterium *Acinetobacter* into the Harbor-UCLA intensive care unit . . . the progenitor of the pan-resistant (meaning resistant to all available antibiotics) *Acinetobacter* that ultimately killed Mrs. B. from chapter 1. I remember Mr. A. distinctly, because I was there, consulting on his case, shortly after he arrived.

When I first met Mr. A., I had finished my three-year residency in internal medicine, and I was three-quarters of the way through my fellowship, training to subspecialize in infectious diseases. For the first eighteen months of my fellowship, I think I may have seen one case of *Acinetobacter*. Harbor-UCLA has one of the most developed antibiotic stewardship programs in the country. Under this program, physicians at Harbor-UCLA can only prescribe powerful antibiotics if they get permission from an infectious-diseases specialist. Our goal is to make sure that the most powerful antibiotics are used only when absolutely necessary.

It is a real burden on physicians at the hospital to have to ask for permission to use an antibiotic. It is also a major burden on the infectious-diseases fellows getting paged literally fifty or a hundred times per day by physicians all over the hospital who are seeking to use these powerful antibiotics. But there is a very good reason for us tolerating these burdens. By ensuring that powerful antibiotics are used only when necessary, and that less-powerful antibiotics are used when they will be effective, we aim to minimize use of the most precious antibiotics, thereby slowing the spread of bacterial resistance to those antibiotics. For many years Harbor-UCLA saw very few drug-resistant infections. *Acinetobacter* was an incredibly rare bacterium at our hospital. Until Mr. A.

As I mentioned in the last chapter, if there is a bright side to the MRSA story, it's that new antibiotics have become available to treat MRSA in the last few years. As a result, even more concerning than MRSA infections are infections caused by other types of bacteria, for which there are no new drugs coming out. Tuberculosis is an important example. Again, you may be surprised to learn that tuberculosis is a still a problem in the twenty-first century. "Isn't TB a thing of the past?" I often hear people ask. The answer is, most definitely not! In fact, there are almost nine million new cases of tuberculosis throughout the world each year,[1] and believe it or not, there were more than 14,000 new cases of tuberculosis diagnosed in the United States in 2005.[2]

Not only are these infections still occurring all over the world, they are becoming more and more difficult to treat. Drug resistance among tuberculosis strains is rampant,[3] and the frequency of drug-resistant cases is increasing not just in underdeveloped countries, but also in countries with advanced medical technology, including the United States.[4] Most alarming are recent descriptions of so-called extreme drug resistant, or "XDR" tuberculosis.[5] XDR strains are resistant to most or all known anti-TB drugs.

These XDR-TB cases threaten to take us back to the era of tuberculosis sanitaria, where doctors performed barbaric treatments like intentionally jabbing a large needle into patients' lungs to cause the lungs to collapse, thereby starving the bacteria of oxygen. In the era prior to antibiotics, that is precisely how TB was treated. TB requires a relatively high amount of oxygen for growth, and humans can survive even with only one lung functioning. So, the idea was to intentionally starve the bacteria of oxygen by—believe it or not—intentionally causing a patient's lung to collapse, thereby hurting the bacteria more than the patient. The idea behind intentionally collapsing the lung to treat infection dates back to about 1820, and it began to be used as a regular treatment for tuberculosis in the late nineteenth century. By 1937, one notable specialist wrote, "At the present time, artificial pneumothorax [i.e., lung collapse] is performed in most institutions for the treatment of pulmonary tuberculosis. In the United States it is probably practiced in a larger percentage of cases treated than all other methods combined."[6]

So, what do you think? If I were to ask you for permission to jab a big needle through your chest or back and into your lung to cause it to collapse like popping a balloon, how fast would you sign up? You may think that this barbaric ritual is but a thing of the distant past—the dark ages of medicine—and that no sane person would consider such a crude and brutal therapeutic procedure in the twenty-first century. If so, you might be interested in the following research article: "Reviving an old idea: Can artificial pneumothorax play a role in the modern management of tuberculosis?"[7] It was published in 2006. The conclusion of the study? Intentional lung collapse "can be considered a useful addition in managing certain patients with cavitary TB, particularly those with drug resistance."

We need to accept the idea that antibiotic resistance is a phenomenon that threatens to send us back to the dark ages of health and medicine. While XDR-tuberculosis is mostly an international problem at the current time, affecting primarily Africa, Eastern Europe (in the countries of the former Soviet Union), and Latin America, it is inevitable that such strains will spread to the United States. In fact, they are already here, albeit in small numbers to date.[8]

A recent single case of what was initially thought to be XDR-TB highlights the public-health nightmare these infections represent. A man living in Georgia had active tuberculosis and was planning on flying to Europe on a personal trip. According to news reports, his physician warned him not to travel because he was known to have TB that was at least partially drug resistant, and therefore had the potential to spread a dangerous infection to other people on the plane.[9] However, the news report continued, the patient "disregarded his doctors' recommendation that he not travel [because he] had compelling reasons for traveling."[10] These compelling reasons turned out to be that he had planned to have his wedding in Greece, and simply didn't want to reschedule it. After his wedding, he traveled across Europe, which included four separate flights for his honeymoon. While he was in Rome, continuing his vacation, the laboratory identified his strain as being XDR-TB (mistakenly so—it was later discovered the strain was indeed multi-drug-resistant, but not to the level of XDR). The US CDC therefore contacted him and told him "in no uncertain terms not to take a flight back."[11] He promptly ignored that warning and "he and his wife decided to sneak back into the U.S. via Canada."[12] So he flew back to Canada, and then drove across the border into the United States. He was then placed in mandatory quarantine, the first time such a measure has been utilized by the federal government since 1963. The ultimate irony? He was a personal injury attorney!

So, according to news reports, here we have an educated, professional gentleman who was twice warned not to fly commercially lest he expose passengers and crew to his active TB. He ignored those warnings on no less than six separate occasions, even after being informed that he had a rare and extremely resistant form of the disease. What is most alarming is that this scenario is not unique and does not surprise me in the slightest. I recently took care of a patient at Harbor-UCLA Medical Center who had been diagnosed with active TB in his lungs in Bogotá, Colombia, and who was actively coughing from his infection. His physician in Colombia told this patient to get on a plane to go to the United States for treatment. Never fear though, because the physician told him to cough into a handkerchief on the plane to avoid infecting other people (I'm *not* making this up!). How stupid was this physician's advice? Well, how reassured would you be if the guy sitting next to you on an airplane said, "Don't worry about my cough. Yes, I do have active tuberculosis in my lung, and yes, when I cough I do spray infectious TB particles through the air. But it's okay because I'm trying to cough into my handkerchief each time. If I forgot once or twice, do me a favor and remind me."

These kinds of scenarios play out all the time, under the radar screen of the news media. Every now and then the stories get picked up. In January of 2008, Reuter's Health reported that a woman who had been diagnosed with multi-drug-resistant tuberculosis in India had chosen to fly from New Delhi to Chicago, and then to California, despite being aware of her diagnosis.[13] She presented to a hospital in California with symptoms that included "coughing up blood, fever, and . . . chest pain." Officials at the US Centers for Disease Control commented rather blithely that there was "a potential for transmission of drug-resistant TB to others." Yeah, no kidding.

In light of the fact that people who know they have TB are willing to fly and expose other people to their infection, how are we supposed to prevent exposures caused by the myriad other travelers who may have TB and don't even know it? XDR-TB is already here, and more cases—many more cases—are coming, whether we like it or not. You will be at risk for them when they are imported from abroad, and if you catch XDR-tuberculosis, we will have few, if any, antibiotics with which to treat you. The mortality rate of XDR-tuberculosis cases diagnosed in other countries (admittedly countries with less-advanced medical technology) has been astronomical (i.e., greater than 90 percent).[14] Even in countries with advanced medical technology, and in patients whose immune systems were functioning normally prior to infection, treatment of infections caused by XDR-TB failed 40–50 percent of the time despite maximal medical therapy.[15] So, it is an understatement to say we have a problem with antibiotic-resistant TB, and that we could really, really use some new antibiotics with which to treat these cases.

Finally, we must discuss the biggest problem of all. That's right. Believe it or not, MRSA and XDR-tuberculosis are not the biggest current antibiotic-resistant threats to countries with advanced medical technology. The biggest threats society is dealing with currently are multi-drug-resistant, and increasingly truly pan-resistant, nosocomial (meaning hospital-acquired), gram negative rods.[16] What, you ask, is a gram negative rod? Well the classical Gram stain allows the laboratory to stain bacteria either purple-blue (gram positive) or red (gram negative). Many bacteria come in the shape of either cocci (spherical) or rods. It is the bacteria that are rod shaped and that stain red (i.e., negative) with a Gram stain that are the biggest antibiotic-resistant problems right now.[17]

Gram negative rods are becoming resistant to all known antibiotics. The leaders in this regard are the nonfermenting gram negative rods (which are so called because they don't ferment glucose as an energy source when grown in the laboratory), such as *Pseudomonas*, *Acinetobacter*, and *Stenotrophomonas*. These three organisms are famous for multi-drug-resistance and for developing new resistance even in the middle of a course of antibiotic therapy.

But a variety of other gram negative rods, such as *E. coli*, *Klebsiella*, *Serratia*, *Enterobacter*, and so on, are also increasingly antibiotic resistant. For example, resistance of *E. coli* to oral fluoroquinolone antibiotics (such as ciprofloxacin) is making it increasingly difficult to treat urinary tract infections with oral antibiotics. Unfortunately *E. coli* is by far the most common cause of urinary tract infections, accounting for 80 percent or more of them. As the resistance to fluoroquinolones begins to approach 50 percent in community strains of *E. coli*, it is possible that tens of thousands of women per year will have to be hospitalized to receive intravenous antibiotics just to treat urinary tract infections! How close are we to this nightmare scenario? A recent international study has reported that up to 30 percent of *E. coli* isolates from urinary tract infections are now resistant to fluoroquinolones.[18] That compares to less than 5 percent only a decade earlier. Although the resistance rates are lower for community-acquired infections as compared to hospital-acquired infections, the resistance rate among community-acquired *E. coli* infections has still risen fiftyfold in just the last decade.[19] It is expected that a continued increase is going to occur in the coming decade, which means we may reach 50 percent resistance of community-acquired *E. coli* to fluoroquinolones within the next ten to twenty years.

In New York City there has been a significant outbreak of infections caused by pan-resistant *Klebsiella* strains that have infected patients already hospitalized for other illnesses.[20] These bacterial strains are literally resistant to every antibiotic on the planet. Nearly 50 percent of patients with these bacteria in their blood have died from the infection. Lately, these infections have begun to spread down the eastern seaboard and into the Midwest. Even worse, we have seen a couple of similar strains in Los Angeles in just the last year. So, these strains appear to have spread across the entire United States.

Pan-resistant infections caused by *Pseudomonas* or *Acinetobacter* are even more common and widespread, with similar horrific outcomes.[21] Not only do such multi-drug-resistant bacteria cause thousands of infections per year in the United States, but there are no new antibiotics in the pipeline that have the potential to treat such infections.[22]

Which brings us full circle back to Mr. A., the patient I introduced at the beginning of this chapter. Mr. A. was the patient who brought *Acinetobacter* into my hospital's intensive care units.

Mr. A. was a fiftyish-year-old Caucasian man with a spinal injury. In fact, like the late Christopher Reeve, Mr. A.'s wound was high up in the cervical spine and caused paralysis in all four limbs. After his injury, he was admitted to the nearest community hospital, where he was put on a mechanical ventilator and admitted to the ICU. As I recall, Mr. A. received care at the community hospital for a couple of weeks, until his condition stabilized. At that point, based on a negative wallet biopsy (i.e., he had no insurance), his hospital contacted Harbor-UCLA Medical Center to arrange a transfer. Our hospital had no choice but to accept the patient; we are not allowed to refuse such transfers.

When Mr. A. arrived in our ICU, the surgery team that would be caring for him immediately knew something was wrong. They noted on his chest X-ray that Mr. A. had an extensive pneumonia. They put him on broad-spectrum antibiotics and sent sputum for culture. Of course, his sputum grew *Acinetobacter*. The surgeons called the infectious-diseases team for help and placed the patient in contact isolation to try to prevent the spread of the organism.

When I first met Mr. A., he was a withered man appearing far older than his stated age. His face was cachectic—that is, wasted, like a starving man might appear—and his limbs were shriveled from being bed bound for several weeks, and being in a catabolic state (burning more calories than he consumed) due to his injury and subsequent infections. He was sweaty, disheveled, and unshaven, and had mucus pooling at the back of his throat because of difficulty swallowing properly. He was an infectious-diseases specialist's nightmare, at risk for every conceivable form of hospital-acquired infection, from bloodstream infections caused by his central venous catheter, to pneumonias caused by accidentally inhaling his own oral secretions, to infected bed-sores.

Mr. A. could respond to verbal communications by blinking and could slightly turn his head from side to side. But he had no limb movement. Because of the location of his spinal injury, he had lost the ability to move his respiratory muscles (e.g., his diaphragm, and the muscles between his ribs), so he could not breathe on his own. Hence, he had to be on a mechanical ventilator permanently, with no hope of ever coming off of it. A hole had been cut into the front of his neck (a tracheostomy) to allow for this permanent mechanical ventilation.

In order to examine Mr. A., I had to don a disposable yellow gown and gloves. I had to listen to his heart and lungs with a dedicated stethoscope, which had been left in the room so that resistant bacteria would not be tracked from him to other patients in the ICU. All of this is normal procedure for a patient in contact isolation. One of the first things I noted on physical examination were the torturous signs of modern medicine tracking up and down Mr. A.'s bruised and battered arms. After multiple blood draws per day, throughout the day and night, for more than two weeks, he looked like he had donated his arms to a vampire convention. Meanwhile, behind me, periodically the mechanical ventilator would ring an alarm for a moment, as Mr. A. fought the machine. It can be very uncomfortable to be mechanically ventilated. What if the machine breathes when you are not ready, or even worse, doesn't breathe when you are ready? Mr. A. was totally at the mercy of the machine. He was completely unable to initiate a breath without it.

In this hellish setting, this paralyzed man was confronting the reality of being debilitated in a modern intensive care unit. Meanwhile, we had to try to come up with an antibiotic regimen with which to treat his infection just to keep him alive. Unfortunately Mr. A.'s *Acinetobacter* was particularly resistant, and imipenem was the only drug left to which the strain was susceptible. So, having no other choice, we started imipenem.

Surprisingly, within a few days, Mr. A. started to respond to the imipenem. His infection started to get better. Despite my own dire expectations, Mr. A.'s fever and blood tests were improving. Some time later, I got called by the surgery team. It seemed another of their patients, a certain Mr. E., who was located several beds away, had started to grow *Acinetobacter* from a wound. You can see where this is headed.

Mr. E. was the victim of a car accident. He had broken both of his hips and had an unstable spine, so, like Mr. A., he was bed bound in the ICU. We almost never saw *Acinetobacter* before Mr. A. arrived in our hospital, and within a couple of weeks of his arrival, a second case appeared in the same ICU. Furthermore, the bacteria causing the second case of *Acinetobacter* had the same antibiotic-resistance profile as the first case. That similar susceptibility profile is like a smoking gun. I'm virtually certain that Mr. A. was indeed the source that brought multi-drug-resistant *Acinetobacter* into my hospital, although it is also true that importation of additional strains from other hospitals probably subsequently occurred.

Even if Mr. A. hadn't have been the one to bring the *Acinetobacter* into my hospital, it certainly would have been someone else shortly thereafter. The spread from hospital to hospital was inevitable given the frequency with which *Acinetobacter* is being seen at hospitals throughout Los Angeles, the United States, and the entire developed world.

Unfortunately the *Acinetobacter* infection Mr. E. acquired prevented him from being able to receive surgery to fix his broken legs. So he lay in the ICU, day after day, week after week, fighting off a horribly drug-resistant infection, with broken legs that could not be fixed in the operating room.

As I said, we had placed Mr. A. in contact isolation. Everyone entering his room had to wear disposable gown and gloves. We had a dedicated stethoscope in that room, so doctors didn't transmit bacteria by touching Mr. A. with their own stethoscopes. We had hand washing instructions on the door of the room. The isolation was all done correctly. But, the bacterium spread nonetheless, as happens so frequently every day in hospitals all over the world, despite the best infection-control efforts.

Meanwhile Mr. A. began spiking fevers again after about two weeks of imipenem therapy. Suddenly his heart stopped beating. His nurse called a code blue. Miraculously, the code team was able to resuscitate Mr. A. His chest X-ray showed new pneumonia, on top of his old pneumonia. His lung had formed a "cavity," meaning the infection had literally eroded a hole in Mr. A.'s lung. We re-sent cultures of Mr. A.'s sputum. Not only did the *Acinetobacter* grow again (still susceptible to imipenem, thankfully), but Mr. A. had also developed infection by a second deadly bacterium, *Stenotrophomonas*, which was itself resistant to most antibiotics, including, of course, imipenem. Mr. A. spent more than two months withering away in the intensive care unit, despite the most powerful antibiotics available, the maximal effort of his doctors and nurses, and hundreds of thousands of dollars in healthcare expenditures. Eventually his infection won, despite all of our efforts.

The car accident victim, Mr. E., ultimately lived, but he struggled with his *Acinetobacter* infection for several months in the intensive care unit (again at the cost of hundreds of thousands of dollars). Due to the toll the infection took on him, he will never be the same again.

But the *Acinetobacter* was not done. As has happened at so many medical institutions all over the world, in succeeding months, it spread to every ICU in our hospital. And it's been there ever since. Our infectious-diseases fellows who are now in training think that our hospital has always been plagued by *Acinetobacter*. But I, and my more senior colleagues, know differently. We remember a time, just a few short years ago, when we almost never saw *Acinetobacter*. I wonder if we will ever be able to go back to that place. Or, are we now entering a new era in medicine, a "postantibiotic era," whose dawn is breaking as dusk settles over the short-lived era of effective antibiotics?[23]

The only new antibiotic that can be used to treat resistant gram negative rod infections and that has become available in the past decade is tigecycline. Tigecycline has been a miracle drug for our hospital, as for many others. Developed by the giant pharmaceutical company Wyeth and approved for use in humans by the US Food and Drug Administration (FDA) in 2005 (long after Mr. A. died), tigecycline has activity against multi-drug-resistant *Acinetobacter*. At our hospital, we are now seeing infections on a regular basis caused by *Acinetobacter* that is resistant to every antibiotic except for tigecycline, and possibly to an ancient drug, colistin, which fell out of use in the 1970s because it was so toxic. If we did not have tigecycline, these infections would be essentially untreatable.

The bad news is that we are starting to see tigecycline and colistin resistance in our *Acinetobacter* isolates. We are very restrictive about how we use tigecycline. Even though the drug can be used for a wide variety of infections, we work hard to try to save the drug, only using it for multi-drug-resistant *Acinetobacter* infections (much to the chagrin of Wyeth's sales team!). Yet, despite all of our efforts, resistance is occurring. And we are not alone. Tigecycline resistance started to become widespread within two years of the drug's availability.[24] Indeed, on a recent trip to visit a hospital on the East Coast, I was told that nearly all of the hospital's *Acinetobacter* is already fully resistant to tigecycline. You're probably starting to see the pattern with which infectious-diseases physicians have become so familiar: resistance is inevitable.

The other bad news is that tigecycline has no reliable activity against the other major hospital-acquired gram negative pathogen, *Pseudomonas*. Every year, there are between 200,000 to 400,000 hospital-acquired infections caused by *Pseudomonas* in the United States, and similar numbers of such infections in Europe.[25] For more than a decade, there have been no new antibiotics developed that can treat *Pseudomonas* that is already resistant to current antibiotics. And none are on the horizon.

Infections caused by these and other antibiotic-resistant microbes impact patients as well as clinicians practicing in every field of medicine. These infections incur tens of billions of dollars in annual healthcare costs, and kill or maim hundreds of thousands of Americans and people across the globe.[26] Given their breadth of effect and significant impact on morbidity and mortality, multi-drug-resistant microbes are considered a substantial threat to US public health and national security by the National Academy of Science's Institute of Medicine[27] as well as by the Infectious Diseases Society of America (IDSA).[28]

Aside from their impact on civilian populations, antibiotic-resistant infections also pose a national security threat because of their proclivity for striking military personnel. Infection of combat wounds is a major problem on the modern battlefield, resulting in death, loss of limb, and long recovery times for many injured soldiers.[29] Up to half of combat-related injuries that have occurred in Iraq or Afghanistan during the current war have been infected at the time soldiers arrive at forward-deployed medical facilities.[30] These infections are often caused by MRSA. Indeed, MRSA has been cultured from wounds even at their initial presentation to the field hospital.[31] During subsequent care, up to 50 percent of battlefield wounds treated at rear-echelon hospitals are infected, typically with highly drug-resistant pathogens, such as MRSA, vancomycin-resistant *Enterococcus*, and multi-drug-resistant *Acinetobacter* and *Pseudomonas*.[32] There are few therapeutic options for such infections, and in an increasing number of cases, the bacteria are resistant to every known antibiotic.

Aside from combat injuries, infections affect military forces in myriad other ways. In fact, disease has been a more common reason for evacuation of troops in Iraq and Afghanistan than has been combat injury.[33] Furthermore, as mentioned, antibiotic-resistant infections, such as those caused by MRSA, also occur frequently in settings of close contact and cramped spaces, such as military barracks or forward-deployment units, resulting in disability and time lost from training activities or forward-deployment activities.

As you can see, the scope of the problem of antibiotic-resistant infections is enormous. We can try to slow down the spread of antibiotic resistance, but that won't do any good for patients that already have these infections. The bottom line is, the only way to deal with the threat of drug-resistant infections among both civilian and military populations is to create new antibiotics that will kill antibiotic-resistant microbes.

Now, you may be thinking, "No problem. I'm sure as the need for new antibiotics increases, companies will develop them so they can make money." If that's what you think, keep reading.

CHAPTER 5
lack of antibiotic development

Several years ago, I received an e-mail from a close friend whom I had trained with during internal medicine residency, Dr. Vivek Bhalla. At the time, Vivek was completing his fellowship in nephrology (kidney specialist) at the University of California, San Francisco. It seemed that Vivek had a friend of a friend who had gotten into a severe motor vehicle accident. This Mr. F. was a young, otherwise healthy man, in his early thirties. Unfortunately, Mr. F.'s injuries from his car accident were quite severe, and he was being cared for in the intensive care unit at an outside hospital (i.e., not my hospital, and not Dr. Bhalla's hospital). Among Mr. F.'s many injuries was severe damage to his chest, which had led to the development of an *empyema*. An empyema is a collection of infected pus inside the chest, trapped between the lung and the chest wall. Empyemas are very difficult to treat because antibiotics don't get into pus-filled spaces very well, and don't get into the area between the lung and the chest wall (the pleural space) very well.

Mr. F. had received numerous antibiotics over the preceding weeks, to treat a variety of infectious complications of his injuries. Despite all these antibiotics—or rather, because of all of these antibiotics—Mr. F. most recently had become infected with a truly pan-resistant *Pseudomonas* strain. The strain causing Mr. F.'s empyema was resistant to every antibiotic available. What could be done?

In light of the explosion across the globe of antibiotic-resistant infections, you might think that pharmaceutical companies would be busy working away to discover and develop new antibiotics with which to treat these infections. But you'd be wrong.

For many years now, leading members of the Infectious Diseases Society of America (IDSA) have been aware that antibiotics were no longer being developed by many pharmaceutical companies. Indeed, many pharmaceutical companies have actually completely eliminated their research and development programs for antibiotic drug discovery.[1]

Although a general decline in antibiotic development had long been appreciated,[2] the initial public alert to the medical community about the true extent of the current crisis in antibiotic development was published by Drs. David Schlaes and Robert Moellering in 2002,[3] in the IDSA's journal, *Clinical Infectious Diseases*. David Schlaes, MD, PhD, a former vice president at the pharmaceutical giant Wyeth, and then at the biotechnology company Idenix, is a pharmaceutical insider with direct and specific knowledge of the abandonment of antibiotic research and development. His coauthor on the warning letter, Dr. Robert Moellering, is a world-famous academic infectious-diseases specialist. Dr. Moellering is the Shields Warren-Mallinckrodt Professor of Medical Research at Harvard Medical School, former physician-in-chief and chairman of the Department of Medicine at Beth Israel Deaconess Medical Center in Boston, and past president of the IDSA. Given his enormous contributions to academic medicine in publications, combined with his prestigious academic position, Dr. Moellering's coauthorship of this warning letter added considerable gravitas.

Subsequent letters from other leaders in medicine confirmed the problem.[4] One of the succeeding letters was from Steven Projan, PhD.[5] Dr. Projan (formerly vice president of biological technologies at Wyeth; currently global head of infectious diseases at Novartis) is another world-famous industry expert with tremendous drug-development expertise. In fact, Dr. Projan spearheaded the discovery and development of the lifesaving new antibiotic tigecycline. As you might guess, the collective voice of these distinguished individuals, who were calling out a warning about lack of antibiotic development, added tremendous credibility to the concern that this problem was severe and imminent.

In response to such publications, as well as to considerable discussion and communications between the IDSA and pharmaceutical and biotechnology officials, several critically important meetings were held to better understand the barriers to antibiotic research and development.[6] The organizations participating in these meetings included (1) the IDSA; (2) the US National Institutes of Health (NIH), which is the single largest government organization funding biomedical research in the world, is funded by US taxpayer dollars, and funds and oversees biomedical research by scientists all over the country; (3) the US Food and Drug Administration (FDA); (4) the Pharmaceutical Research and Manufacturer's Association of America (PhRMA), which is the trade organization of the pharmaceutical industry; and (5) the Biotechnology Industry Organization, which is the trade organization of the biotechnology industry. The purpose of these meetings was to bring together experts from all aspects of antibiotic development and infectious diseases to discuss what had gone wrong with antibiotic development. While these meetings generally led to a better understanding of the causes of the lack of antibiotic development (more on this to come), few actionable items were generated to allow solutions to be created. Subsequently, the IDSA established a specific task force, the Antimicrobial Availability Task Force (AATF), to consider options, develop recommendations for legislative and administrative action, and raise public awareness about the problem.

The AATF is composed of a highly distinguished group of internationally known infectious-diseases specialists. The task force is chaired by Dr. John Bartlett (Stanhope Bayne-Jones Professor of Medicine and previous chief of the division of infectious diseases at Johns Hopkins University School of Medicine, as well as past president of the IDSA), and includes such other distinguished members as Dr. David Gilbert (past president of the IDSA, professor of medicine and chief of infectious diseases at Providence Portland Medical Center and the Oregon Health

Sciences University, and the editor of one of the most widely used professional handbooks in all of medicine, *The Sanford Guide to Antibiotics*), Dr. W. Michael Scheld (Bayer-Gerald L. Mandell Professor of Internal Medicine at the University of Virginia and past president of the IDSA), Dr. John Bradley (director of the division of infectious diseases at the Children's Hospital in San Diego, Skaggs Research Scholar at the Scripps Research Institute, and associate professor of pediatrics at the University of California, San Diego), Dr. Helen Boucher (assistant professor of medicine and director of the infectious-disease training program at Tufts New England Medical Center), Dr. John Edwards (chief of the division of infectious diseases at my hospital, Harbor-UCLA Medical Center, and professor of medicine at the Geffen School of Medicine at UCLA), Dr. George Talbot (an infectious-diseases specialist and consultant with considerable drug-development expertise in both small biotechnology companies and large pharmaceutical companies), and, most recently, myself (I joined in 2007). Finally, Bob Guidos, Esq., is the IDSA staff member who provides policy and legislative guidance and organization to the AATF.

The AATF was charged to attack the problem of disappearing anti-infective discovery and development. The task force was funded (and continues to be funded) solely by financing from IDSA member dues, with no money accepted from pharmaceutical companies. The AATF began its offensive between 2003 and 2004 (prior to my joining the task force). It began by stumping Capitol Hill, trying to educate congressional staffers and congresspersons, and trying to rally support for the idea that not having antibiotics is bad, and that finding ways to stimulate antibiotic development would be good.

The AATF hit a brick wall very quickly. Congresspeople wanted hard evidence that pharmaceutical companies were indeed abandoning antibiotic development. Where were the data? Newspaper stories? Rumors? Personal communications? Those weren't cutting it. The AATF needed some sort of hard data proving that anti-infective development was declining.

In 2004, I was asked by Dr. Edwards to help the AATF accumulate such data. We collaborated with John Powers, MD, who at the time was the lead medical officer for antimicrobial drug development and resistance at the FDA, and several other experts, including Dr. Eric Brass, who is the past chairman of the FDA Nonprescription Drugs Advisory Committee and past chairman of the department of medicine at Harbor-UCLA Medical Center. Together we sifted through FDA databases on newly approved drugs going back twenty years. I also sifted through the annual reports of the fifteen largest pharmaceutical companies in the world, and the ten largest biotechnology companies to understand better what the drug pipeline for the future held. My colleague, Loren Miller, MD MPH, performed an independent audit of my findings to eliminate any errors, omissions, or misclassifications. What we found was very surprising.

We published our results in 2004.[7] Our study provided the AATF with the first peer-reviewed data confirming the decline in development of new antibiotics. The most important finding was that, as of 2004, there had been a 67 percent reduction in new antibiotics that became available for use in the United States between the years 1983 and 2002. That trend showed no sign of reversing through 2008, and in fact has continued to worsen, as evidenced by an updated chart of the same data now showing a 75 percent decline in antibiotic development since 1983 (fig. 5.1). This figure leaves absolutely no doubt. Antibiotic development was and is deteriorating at an alarming rate.

My anticipation is that over the course of the current five-year period (2008–2012), the number of new antibiotics approved for use in the United States will plateau. That is, unless specific changes are made in drug development and regulatory review of drugs in development, I anticipate something on the order of five to seven new antibiotics to be approved in the United States over that time frame. Even worse is that none of those new antibiotics are likely to be able to treat bacteria that are resistant to similar antibiotics that are already available in the United States. So, how much does it help us to have new drugs developed that are no better at treating bacteria that are already resistant to the drugs we already have? Some of these new drugs may achieve incremental improvements compared to currently available drugs. For example, they may cause fewer side effects than drugs currently on the market. Some of the

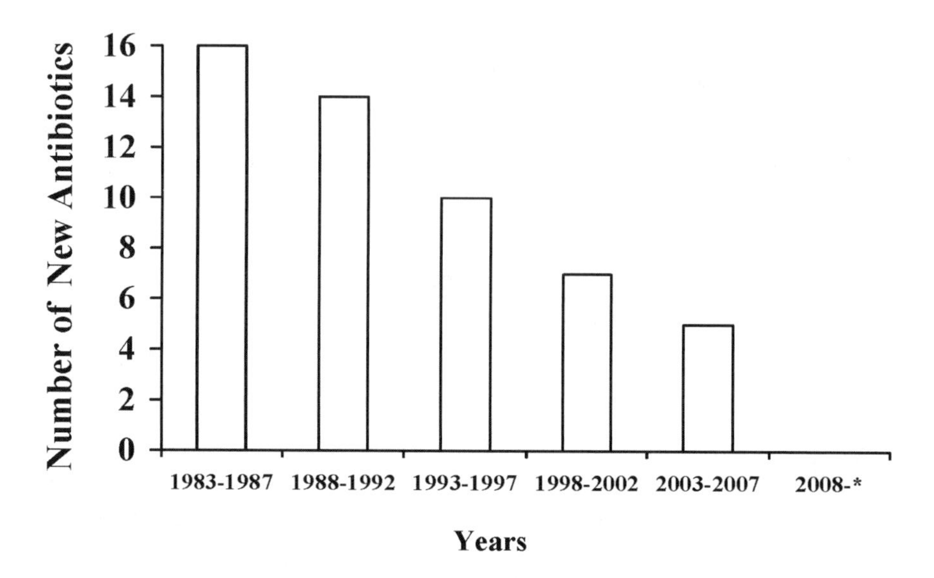

Figure 5.1. Antibacterial Drugs Approved by the FDA for Use in Humans per Five-Year Period. This graph shows the alarming and incontrovertible decline in the availability of new antibiotics in the United States over the last 25 years. Updated from Spellberg et al., "The epidemic of antibiotic-resistant infections: A call to action for the medical community from the Infectious Diseases Society of America," *Clinical Infectious Diseases* 46 (2008): 155–64. Data are up to date through the first half of 2009.

new drugs may be oral drugs, as compared to current drugs that can only be given intravenously. But the bottom line is, what we need immediately and in the future are new drugs that can treat bacteria resistant to drugs that are already available.

Furthermore, it takes on average well over ten years from the time an antibiotic is discovered until it is approved for use in patients.[8] Therefore, if society were to somehow magically fix this problem, and new antibiotic development were to be reinvigorated tomorrow, it would not be until the year 2019 at the earliest that we could expect this graph to truly start going in the other direction as new antibiotics became available. This is what we, in the field, refer to as "bad."

In 2004, when we looked at the publicly available listings of the product pipelines of the world's fifteen largest pharmaceutical companies, we found only five new antibacterial drugs listed as being in development. The five antibiotics in development were actually fewer than the eight drugs in development to treat bladder hyperactivity and the seven drugs in development to treat either acid reflux or irritable bowel syndrome. Furthermore, the five antibiotics in development barely beat out the four new drugs in development to treat erectile dysfunction. Is that a comment on our society or what? I mean, maybe it's just me, but it seems like not having antibiotics with which to treat life-threatening infections is perhaps somewhat more serious of a societal problem than is an insufficient number of erections in the United States and throughout the world!

Some people have suggested that as large pharmaceutical companies withdraw from antibiotic development, biotechnology companies will fill the gap. However, in 2004 we found only one new systemic antibacterial agent listed as being in development by the world's largest biotechnology companies. A recent follow-up study conducted by the AATF reaffirmed the ongoing dearth of antibiotic development not only by big pharmaceutical companies but also by smaller biotechnology companies.[9] Other surveys have concluded the same.[10]

The lack of antibiotic development is not the result of declining investment by pharmaceutical companies in their overall research and development portfolios. Indeed, we found that between 1998 and 2002, the ten largest pharmaceutical companies increased their collective research and development budget by 31 percent, from $21.9 billion to $28.6 billion per year.[11] Furthermore, the Pharmaceutical Research and Manufacturer's Association (PhRMA) has indicated that its members' collective research and development budgets increased by more than 70 percent between the years 2000 and 2007, going from $26 billion to $44.5 billion (fig. 5.2).[12] Indeed, going back further, pharmaceutical company research and development investments increased an amazing 430 percent between 1990 and 2007, and an even more staggering 2,125 percent from 1980 to 2007. These numbers reflect an astounding average annual increase of 79 percent in research and development spending over the last twenty-seven years. So, the decline in development of antibiotics is specific to research and development of antibiotics relative to other

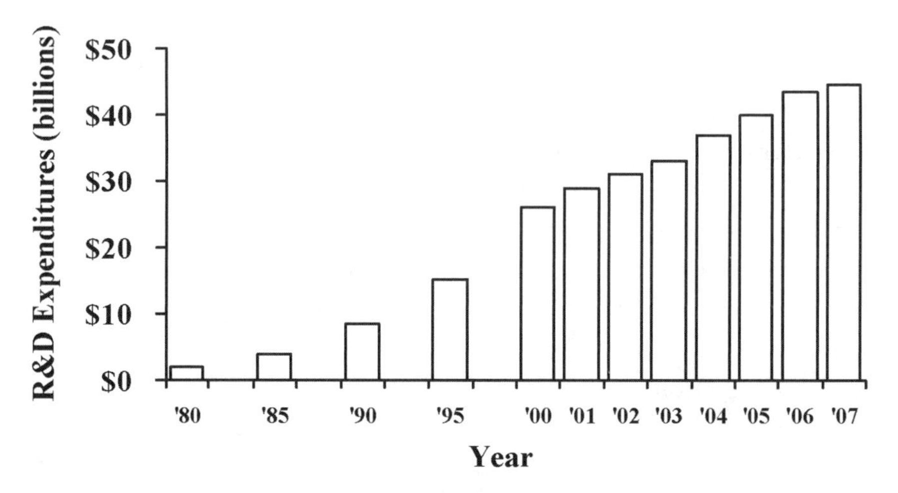

Figure 5.2. Collective Annual Research and Development Expenditures by Companies with Membership in the Pharmaceutical Research and Manufacturers of America (PhRMA). Reprinted with permission from PhRMA.

classes of drugs, and is not the result of a general decline in research and development expenditures at pharmaceutical companies.

In concordance with industry data on expenditures for research and development, a recent academic analysis of clinical trials conducted in all areas of medicine found that the number of clinical trials in infectious diseases, and in particular, for bacterial infections, is declining.[13] In contrast, the number of clinical trials in other medical areas, such as cancer, arthritis, cardiovascular diseases, and so on, is continuing to increase. Again, the decline in research and development of new antibiotics is not indicative of a general decline in pharmaceutical company research and development in all areas. Antibiotic development—far more so than drug development in general—is dying.

Following publication of our manuscript quantifying the decline in antibiotic development, in July 2004, the IDSA released a report: "Bad

Bugs, No Drugs: As Antibiotic Discovery Stagnates, a Public Health Crisis Brews."[14] The IDSA used this report to try to publicize the problem of lack of antibiotic development, and to further educate congresspeople about the problem and about solutions to it proposed by the IDSA. Once again, the IDSA financed this advocacy campaign with patients' best interests and the public's health in mind. No funding from pharmaceutical sources was accepted for this effort.

Subsequently, IDSA leaders have testified at multiple governmental and congressional-level hearings,[15] have been interviewed for a significant number of trade and major news stories, and have published op-ed pieces to try to further publicize the problem.[16] The IDSA also actively participated with the public television program *Nova*, which produced an Emmy award–winning episode, "Rise of the Superbugs," which aired in the winter of 2005. The "Rise of the Superbugs" program told the story of epidemic antibiotic resistance using patient stories to underscore the point.[17] The show was even narrated by Brad Pitt. Unfortunately, while the episode was critically acclaimed, and despite the highly visible profile of the narrator, the public did not seem to notice and the program seems to have had little meaningful impact on general awareness of the problem. Most recently, *Time* magazine has acknowledged the problems of increasing drug resistance and lack of new antibiotic development, as well as the role of the IDSA in lobbying Congress to address these problems.[18]

Given this intense activity to publicize the problem of lack of antibiotic development, we were initially optimistic that a political solution might be forthcoming. But for unclear reasons, all of our efforts to bring the problem of lack of antibiotic development before the public's attention have failed. While people seem to have become more aware of the problem of bacterial infections recently, they are no more aware today of the lack of antibiotic development than they were five years ago.

In the aftermath of its release of the "Bad Bugs, No Drugs" white paper, the IDSA worked with several members of Congress to create legislation that could have gone a long way toward stimulating antibiotic research and development, and several promising bills were introduced on Capitol Hill. Unfortunately, as a whole, the 109th Congress focused on other priorities and did not act on these bills prior to adjourning in December 2006.

There are recent signs that the federal government may be starting to realize the extent of the problem. In September 2007, both the House and the Senate voted to approve the bipartisan FDA Amendments Act, among the many components of which was reauthorization of the Prescription Drug User Fee Act (PDUFA). The FDA Amendments Act and the PDUFA focus on improving the FDA's ability to carry out its critical safety-monitoring role for drugs, food, and medical devices. Also included are provisions, developed with IDSA guidance, that will enable the government to begin to gather badly needed data about the extent of the spread of antibiotic resistance among bacteria.

Another piece of legislation that was developed with IDSA input, the Strategies to Address Antibiotic Resistance (STAAR) Act, has also been introduced to the House of Representatives in a bipartisan way by Representatives Jim Matheson (D-UT) and Michael Ferguson (R-NJ) and into the Senate by Senators Sherrod Brown (D-OH) and Orrin Hatch (R-UT).[19] The STAAR Act calls for creation of an Office of Antimicrobial Resistance in the Department of Health and Human Services, and a Public Health Antimicrobial Advisory Board, which are intended to develop coordinated plans and manage a federal effort to combat antibiotic-resistant infections. This coordinated effort would include gathering data on how common such infections are and tracking the spread of such infections in real time. These bills are seen as the first concrete, positive steps toward beginning to address the crisis in antibiotic resistance, and they are strongly supported by the IDSA and the infectious-diseases community at large. They have also been endorsed by a number of other professional medical organizations, including the American Medical Association, the American College of Physicians, the American Public Health Association, the Society for Healthcare Epidemiology of America, the Association for Professionals in Infection Control and Epidemiology, the Pediatric Infectious Diseases Society, the National Foundation for Infectious Diseases, the Society of Infectious Diseases Pharmacists, the Alliance for the Prudent Use of Antibiotics, and Premier, an alliance of

1,700 nonprofit hospitals and healthcare systems nationwide. And again, these bills have bipartisan political support, underscoring the universal nature of this issue.

Unfortunately, the STAAR act did not get passed before the end of the 110th Congress. It has recently been reintroduced during the 111th Congress and hopefully will be passed soon. Meanwhile, both the FDA Amendments Act and the STAAR act focus on forces that drive antibiotic resistance among microbes, and neither bill does anything to actually intervene in the problem of lack of antibiotic development. That is, they focus exclusively on prevention of antibiotic resistance, without dealing with the needed research and development of new antibiotics. An analogy to illustrate the difference would be if the government were to focus exclusively on increasing the fuel efficiency of cars without developing alternative, renewable-energy sources to deal with the global-warming crisis. Conservation is critical, but ultimately all conservation does is buy more time to create new resources. If new resources are never created, ultimately conservation only delays the inevitable exhaustion of the resource in question.

Creating a government structure to deal with antibiotic-resistant infections, and gathering data on their frequency and geographical spread, is a critical step in combating these infections from a public-health perspective. However, these features will not enable development of new antibiotics. Nor will they enable treatment of infections that are already resistant to current antibiotics. Thus, while these bills are important and meritorious, they must be accompanied by other efforts to reinvigorate the discovery and development of new antibiotics.

Most recently, legislation sponsored by Senator Charles Schumer (D-NY) was introduced into the Senate and House that will provide tax credits to make up for research and development expenses for products designed to target infectious diseases.[20] This legislation is another promising step forward on the path to correcting our current deficiency in antibiotic research and development. But, while a step in the right direction, research and development tax credits are only a small step, and are likely not enough to pull large pharmaceutical companies back into the antibiotic development arena. What is desperately needed now is a series of large steps to stimulate discovery and development specifically of priority antibiotics (i.e., those that are effective against serious or life-threatening infections that are resistant to available antibiotics).

In the meantime, the IDSA has continued to work with key members of Congress, including Senator Hatch and Senators Edward Kennedy (D-MA), Michael Enzi (R-WY), Richard Burr (R-NC), Sherrod Brown (D-OH), Lamar Alexander (R-TN), Chris Dodd (D-CT), Richard Burr (R-NC), and with members of the US House of Representatives including Representatives Jim Matheson (D-UT), Henry Waxman (D-CA), John Dingell (D-MI), Brian Baird (D-WA), and Barbara Cubin (R-WY), to develop comprehensive legislation to address the burgeoning problem of antibiotic-resistant infections. It should also be specifically emphasized that IDSA previously met with then-Senator Barack Obama's staff, which raises hopes for further accomplishments in the future now that he is president—and a president who has vowed to bolster science, medicine, and technology.

In the meantime, here we are, more than five years after Schlaes and Moellering published their initial warning letter about the crisis in antibiotic development, and after years of the IDSA's effort, and there is still no sign that any solutions for solving the antibiotic crisis will be forthcoming in the next decade.

When Dr. Bhalla e-mailed to ask if I had any suggestions for last-ditch efforts to treat his friend, Mr. F., the car accident victim with the pan-resistant *Pseudomonas* around his lung, I had no brilliant insights to offer. Together we came up with a witch's brew of antibiotics to be used in desperation, despite the fact that the *Pseudomonas* was resistant to them, in the hopes that combining multiple antibiotics would cause their effect to be greater than the sum of their parts. This regimen was the equivalent of tossing eye of newt, toe of frog, and wing of bat into a bubbling cauldron, while dancing around it and rubbing together two chicken bones and a setting fire to a cup of rum.

Mr. F. ended up going to surgery for what is called a "decortication" procedure. In this procedure, Mr. F.'s rib cage was cracked open, and the surgeons literally scraped and scooped out as much pus and infection from Mr. F.'s chest as was possible. Decortication is a down-and-dirty, brutish procedure that has a high complication rate and invariably takes months to recover from, if the patient survives. Fortunately, Mr. F. did survive, but I am told that this thirty-ish-year-old man, previously in good health, was left almost totally incapacitated. Whether or not the witch's brew of antibiotics had any effect is debatable. Certainly Mr. F.'s chances for a meaningful recovery would have been significantly enhanced by having new antibiotics with which to treat his resistant infection.

How bad is the current crisis in antibiotic resistance? There are no new antibiotics in the pipeline that will be available in the next decade to treat *Pseudomonas*, *Acinetobacter*, or other gram-negative bacterial infections that are already resistant to all current options. The situation for anti-TB antibiotics is just about as bleak, while the situation for antibiotics for gram-positive bacterial infections is less grim in the short term, but not the long term.

The real issue is that politicians and policymakers have not yet been convinced that the crisis posed by drug-resistant bacteria and the lack of antibiotic development are of sufficient importance, or have reached a significant enough level of public alarm, to act in a meaningful way. In the meantime, more and more patients are dying of increasingly resistant infections that are being treated with fewer and fewer new drugs.

CHAPTER 6
the war against microbes?

CAUSES OF MICROBIAL RESISTANCE

Part of the problem in convincing people that we need to create new antibiotics stems from a general misunderstanding about the causes of, and therefore the solutions to, the explosion in antibiotic-resistant infections. There is no question that the global spread of bacterial, fungal, and parasite resistance to antibiotics is a predominant reason why infectious diseases are increasingly plaguing humanity, and will continue to do so for the foreseeable future. But why is microbial resistance so prevalent? What causes microbial resistance to occur, and what causes it to spread? Can these causes be reversed?

The single most commonly expressed reason why antibiotic resistance occurs is physician misuse of antibiotics. The thought process goes: "Physician misuse and overuse of antibiotics causes antibiotic resistance among microbes. Therefore, if we could only convince physicians to use antibiotics responsibly, we could stop antibiotic resistance and thereby win the war against microbes." Unfortunately, despite how widespread this belief is, it is nevertheless wrong. Physician misuse of antibiotics does *not* cause microbial resistance. This belief reflects an alarming lack of respect for the incredible power of microbes.

The fact that we can't see microbes causes us to lose perspective about their truly mind-boggling numbers. It is worth considering that, despite being smaller than one millionth of a meter long, microbes comprise fully 60 percent of the mass of life on the planet (90 percent of the living mass if plant cellulose is excluded from the calculation).[1] This astonishing fact reflects the staggering numbers and diversity of microbes. There are an estimated 5×10^{31} microbes living on planet Earth, with an estimated collective mass of 50 quadrillion (5×10^{16}) metric tons (1 metric ton = 1000 kg).[2] By comparison, there are 6 billion (6×10^{9}) human beings on the planet, with an approximate collective mass of 300 million (3×10^{8}) metric tons.

These numbers are so large that they are difficult to fathom. So let me give an illustrative comparison. If you wanted to make the mass of human beings on planet Earth equal the mass of microbes, you would have to take every person on Earth and copy every single one of us 100 million times. Of course, while you were making 100 million copies of every person on Earth, you would inadvertently be copying microbes as well, since enormous amounts of bacteria and fungi live on and in our bodies at all times. Indeed, from the microbial perspective, human beings are nothing more than walking microbial planets. Believe it or not, there are approximately five to ten times more microbes living on and in every human being than there are human cells in our bodies—between 50 and 100 trillion microbes versus about 10 trillion human cells in a human body.[3]

As diverse as human beings are, we pale in comparison to the adaptability of microbes, which inhabit literally every possible clime and environment on the planet, despite extremes of boiling or freezing temperatures, pressures sufficient to crush virtually any human-made submersible or submarine device, extreme salinity, zero oxygen content, presence or absence of sunlight, and so on. Bacteria even exist in large numbers miles deep in the midst of solid rock in the earth's crust.[4]

The power that drives this astonishing degree of adaptability is rooted in two fundamental features of microbial biology. First is the ability of bacteria to reproduce almost impossibly quickly. It takes many bacteria only twenty to thirty minutes to replicate; it takes human beings twenty to thirty years to replicate. Think about that replication time difference for a moment. By reproducing every twenty minutes, a single *E. coli* bacterium can create 69 billion progeny in just twelve hours of logarithmic growth. Now *that's* an enemy for you. Kill one, and as many as 69 billion more can pop out in twelve hours!

The second biological feature of microbes that drives their awesome adaptability is their virtually limitless ability to create diversity in their genes. Each time a bacterium reproduces, it can pass on slight changes in the genetic code of its progeny. These changes are created by errors made in the process of copying the progenitor's DNA. Of every million progeny produced by a single dividing bacterium, up to 30,000 (i.e., 3 percent) may have unique, random DNA mutations. Of course, most of these mutations (more than 99 percent) are either harmful or neutral to the survival of the mutant bacteria. But, a small percentage of the mutations, estimated to be about 0.5 percent, result in enhanced ability of the mutant strain to survive in its current environment.[5] As a result, rapid bacterial reproduction actually encourages genetic diversity. In twelve hours of growth, one bacterium can create 69 billion offspring that are generally very similar to the progenitor. But 230 million of the offspring may contain unique mutations that cause them to differ slightly from their ancestral strain, and 1.5 million of the offspring may contain mutations that specifically enhance their ability to survive despite hostile environmental conditions, such as the presence of antibiotics. The potential variability resulting from this astounding replication and mutation rate is virtually limitless. It is the creation of these unique genes at this unbelievable rate that allows bacteria to adapt quickly to any environment, including hostile environments containing antibiotics that would otherwise be lethal to the bacteria.

Furthermore, unlike humans, bacteria can exchange DNA with neighbors even without reproducing (i.e., within a single generation). Humans can only swap DNA when a male and a female create a child. Not so for bacteria, which can swap DNA with one another even when they are not reproducing. Bacteria can simply reach out and grab one another using sexual appendages called "pili," and pass strands of DNA through the connections. For example, bacteria might swap a penicillin-resistance gene for a gene that allows them to grow at a different temperature, or in a high concentration of salts, or for a resistance gene targeting a different antibiotic. Furthermore, when bacteria trade genes, they may keep copies of both genes involved in the trade, building up huge collections of such genes over eons of time.

Even more amazing is that while humans can mix DNA only within our own species, bacteria are incredibly promiscuous. You may recall from basic biology that the standard way that organisms are classified is in the following categories (in descending order): kingdom (e.g., animal vs. plant vs. bacteria, etc.), phylum (e.g., Chordata [creatures with spinal cords]), class (e.g., Mammalia), order (e.g., Primate), family (e.g., Hominidae), genus (Homo), and species (*Homo sapiens*). Within the entire range of diversity of life on Earth, humans can exchange our DNA only with other members of our own species. In contrast, bacteria can reach out, grab, and mix and match DNA with bacteria of another species, genera, family, order, class, and all the way up through the phyla level.[6]

Let's put this promiscuous spread of genetic material into perspective using an illustrative example. Humans are genus/species *Homo sapiens*, family Hominidae, order Primate, class Mammalia, and phylum Chordata. So, for humans to demonstrate an equivalent degree of genetic promiscuity to that of bacteria (i.e., cross-phylum) would require humans to be capable of exchanging our DNA with a chimpanzee (family Hominidae), an orangutan (order Primate), a grizzly bear or a tiger or a killer whale (now THAT would be something!) or a walrus (all class Mammalia), or a falcon, great white shark, frog, crocodile, or, most bizarre of all, a sea squirt (all of which are in the phylum Chordata). Think of the genetic diversity that would be possible if we could do that, from the monstrous to the magnificent, and everything in between.

This ultrapromiscuous bacterial mixing of DNA has direct implications for the spread of microbial resistance. For example, bacteria in the phylum Actinomycetes long ago evolved to produce a group of antibiotics called the aminoglycosides. The original Actinomycetes inventors presumably used the aminoglycosides to kill off competing bacteria. Aminoglycosides are among the most powerful antibiotics that humans have ever discovered, and until very recently they had been reliably active even against the deadly pathogens *Pseudomonas* and *Acinetobacter*. Of course, since the Actinomycetes bacteria learned how to make aminoglycosides, they also had to learn how to defend themselves from the aminoglycosides so they wouldn't commit suicide with the very aminoglycosides they were producing. So around the same time that the Actinomycetes invented aminoglycosides, they also invented aminoglycoside-resistance genes, and hid these genes on mobile genetic elements called plasmids. Plasmids are like genetic MP3 players, which can contain within them a veritable library of genes all sorted and stored in their appropriate place. These genetic MP3 players allow for easy sharing of these genes with other bacteria—as easy as downloading songs from the Internet.

In the last few years, scientists have discovered that plasmids from the Actinomycetes have transmitted genes to bacteria worldwide, including into members of the phylum Proteobacteria. As a result, the dreaded *Pseudomonas* and *Acinetobacter* (both of which are Proteobacteria) have acquired the genes that enable them to resist aminoglycosides. Because of these shared genes, *Pseudomonas* and *Acinetobacter* are increasingly resistant to aminoglycosides.[7]

The power of microbial genetic plasticity is absolutely awesome. Now, combine this awesome genetic plasticity with the fact that the oldest known microbial fossils are 3.5 billion years old.[8] So, while our species has been around for about 4 million years, bacteria have been adapting to the various environments on planet Earth for one thousand times longer, through innumerable geological cataclysms, ice ages, catastrophic meteor crashes, mass extinctions, and every other form of apocalypse that the Earth has ever undergone. Given their awesome power to create variability in their DNA at an astonishingly rapid pace, and the mind-bogglingly long period of time in which they have had to do so, it should not be surprising that microbes are the most numerous, diverse, and adaptable organisms that have ever lived on the planet.

It is time for some humility on the part of *Homo sapiens*. We must abandon the frequently used metaphor of humans being "at war with microbes."[9] It is absurd to believe that we could ever defeat in a war organisms that outnumber us by a factor of 10^{22}, outweigh us by a factor of 10^{8}, have existed for a thousand times longer than our species, can undergo as many as 500,000 generations during one human generation, and can even create new weapons and defense mechanisms within generations (table 6.1).

Furthermore, the weapons we would use in a "war with microbes" would be antibiotics. We need to remember that human beings did not invent antibiotics, we merely discovered them. Virtually all of the antibiotics we now use are either harvested directly from microbes or are made synthetically based on the design of naturally occurring antibiotics.[10] Genetic analysis of microbial metabolic pathways indicates that microbes first invented both antibiotics and resistance mechanisms to defeat those antibiotics more than two billion years ago.[11] In contrast, antibiotics were not discovered by humans until the first half of the twentieth century. Hence, microbes have had collective experience creating and defeating antibiotics for twenty million times longer than *Homo sapiens* have known antibiotics existed. Indeed, so experienced and successful are bacteria at developing resistance to antibiotics that some have actually evolved to be able to survive by ingesting and using antibiotics as their only food source![12]

Finally, even if by some miracle human beings created a way to

TABLE 6.1. MICROBES VS. HUMANS

	Microbes	**Humans**	**Factor**
Number on Earth	5×10^{31}	6×10^{9}	$\sim 10^{22}$
Mass (metric tons)	5×10^{16}	3×10^{8}	$\sim 10^{8}$
Generation Time	30 min	30 yr	$\sim 5 \times 10^{5}$
Time on Earth (yrs)	3.5×10^{9}	4×10^{6}	$\sim 10^{3}$
Experience with Antibiotics (yrs)	2×10^{9}	<100	$>2 \times 10^{7}$
Exchange DNA across . . .	Phyla	N/A—within species only	

Adapted from Spellberg et al., "The epidemic of antibiotic-resistant infections: A call to action for the medical community from the Infectious Diseases Society of America," *Clinical Infectious Diseases* 46 (2008): 155–64.

utterly and finally defeat microbes in a war, we would only be sealing our own doom. For, when it comes to microbes, it can truly be stated that we can't live with them, and we can't live without them. Microbes provide functions that are critical to the continued survival of humans, and indeed most other life on the planet. For example, bacteria are the only organisms on the planet that are capable of fixing nitrogen from the atmosphere into the food chain.[13] This process is essential to the creation of macromolecules, such as proteins, that are the basis of life. Bacteria also produce a variety of vitamins and cofactors essential to our diet, such as vitamin K.[14]

So forget about being at war with microbes. We are not at war with them, and we should not want to be. We would have no hope of defeating them, and even if we could defeat them, we would only be destroying ourselves. Rather, we must learn how to coexist with microbes, holding at bay those that are harmful, and leaving be those that are helpful.

It is obvious that microbes do not need our help in creating antibiotic resistance. They are doing just fine on their own, thank you very much. On the other hand, what human beings can do is affect the rate of spread of preexisting bacterial resistance by applying selective pressure via exposure to the thousands of metric tons of antibiotics we have used in patients and livestock over the last half century.[15] The use of antibiotics in livestock is particularly problematic, as these antibiotics enter human circulation via food consumption and directly promote the spread of resistant bacteria.[16]

After exposing bacteria to antibiotics, the susceptible bacteria all die, and only the rare bacteria that were already resistant to the antibiotics are left alive. Those surviving antibiotic-resistant bacteria then grow and multiply, spreading their antibiotic resistance. It's a crystal-clear and unequivocal example of natural selection and resulting evolution at work, just as Darwin described 250 years ago.

That antibiotic use can increase the spread of antibiotic resistance is well established. In fact, there is a direct correlation between antibiotic tonnage dumped into the environment and local rates of antibiotic resistance.[17] As I said, this correlation is merely the inevitable outcome of Darwinian natural selection.

As an illustrative, classical example of a similar phenomenon that occurs on the macroscopic level, consider the European peppered and melanic moths (fig. 6.1). These moths live on and around trees throughout Europe and in England in particular. Prior to the Industrial Revolution, almost all the moths in England were of the peppered variety, with scattered spots of black on a white background.[18] Less than 5 percent of the moths were melanic, that is, black all over. Note that the peppered and black moths are actually the same species. The difference in color is due to a genetic mutation that causes some of the moths to overproduce melanin, giving them a diffusely black appearance.

A picture of a moth sitting on a white, lichen-covered tree trunk immediately reveals the advantage of the peppered moth, as compared to

Figure 6.1. Industrial Melanism. Moth images adapted by permission from H. B. D. Kettlewell, "Further selection experiments on industrial melanism in the Lepidoptera," *Heredity* 10 (1956): 287–301, Macmillan Publishers Ltd. (A) Peppered (white) and (B) melanic (black) moths are the same species, just with different colors. During the Industrial Revolution, pollution killed off white-colored tree lichen, and soot covered the tree trunks, making them dark (top panel). The darker color of the tree trunks made peppered moths stand out to predators and allowed melanic moths to hide and escape getting eaten. Therefore, melanic moths increased in frequency and peppered moths declined (Industrial Melanism). In the twentieth century, as the British instituted pollution controls, tree lichen came back, soot covering the trees diminished, and tree trunks again became white (bottom panel). Peppered moths came back and the melanic moths were reduced in frequency.

a melanic moth.[19] The peppered moth is nearly invisible on the white lichen, while the melanic moth stands out like a sore thumb. So, while sitting on tree trunks covered with white lichen, peppered moths are very difficult for predators—mostly birds—to spot, while melanic moths are far easier prey.[20] In pre–Industrial Revolution England, this predator-prey relationship resulted in more of the melanic moths getting eaten by birds, so the vast majority of the surviving moths were peppered.[21]

But an interesting phenomenon happened during the Industrial Revolution. As soot and other environmental pollutants were dumped into the air at an increasing pace, the white-colored lichen that covered the tree trunks began to die off, to be replaced by black-colored soot. The color of the tree trunks therefore began to change from white to black. At the same time, there was a shift in the normal distribution of peppered versus melanic moths, driven by increased predation of the peppered moth and decreased predation of the melanic moths. Since more peppered moths and fewer melanic moths were now being eaten by birds, melanic moths increased in numbers and peppered moths decreased. Hence, pollution dumped into the environment by humans caused a dramatic shift in the genetic background of the moth population in England. This process has been termed "Industrial Melanism," based on the increase in melanic moth frequency that was driven by increased industrial pollution production.

But how do we know that the pollution was actually responsible for causing Industrial Melanism? We have a clear association of higher pollution and more melanic moths. That is, when there was no pollution, there were few melanic moths. When there was more pollution, there were more melanic moths. Does this in fact prove that the increase in pollution *caused* an increase in melanic moths? No. In order to prove that pollution caused the increase in melanic moth frequency, we would have to intervene in the system, reducing the pollution to see if this reversed the increase in melanic moths. Fortunately, in the twentieth century, the British grew wise to the pollution problem, and they passed laws and set standards that resulted in a dramatic decline in airborne soot. As the pollution levels fell, the melanic to peppered moth ratio reversed back to its original state.[22] Now, with the white-colored lichen restored to the tree trunks, since the mid-twentieth century, once again the peppered moth has been dominant, and the melanic moth is again a rarity.

Industrial Melanism is a classic example of Darwinian natural selection in action, and it is a quite compelling illustration of the same principles by which antibiotic use causes shifts in the populations of microbes in the environment. Pouring antibiotics into the environment kills off the susceptible bacteria, leaving behind small numbers of already resistant bacteria. As those resistant bacteria reproduce, and/or as they share their resistance genes with other bacteria that were previously susceptible to antibiotics, the frequency of antibiotic-resistant bacteria increases. The more antibiotics we dump into the environment, the greater the selective pressure—that is, the more rapidly susceptible bacteria will be killed off and replaced by resistant bacteria.

We have to remember that bacteria invented antibiotics to be used in very small quantities acting over very small distances. A single bacterium may produce far less than one trillionth of a gram of antibiotic, and that antibiotic is only intended to attack other bacteria within a millionth of a meter or less around the source bacterium. Microbes have been much more careful in their selective use of these invaluable weapons than have humans. By producing antibiotics in small amounts only in certain circumstances, they have minimized the selective pressure on their neighbors, which allowed their antibiotics to remain widely effective for hundreds of millions, or even billions of years. In contrast, humans mass-produce antibiotics and dump them all over the environment in the thousands of kilograms. As a result, in less than a century, we have caused tremendous shifts in the ecology of bacteria all over the planet. We've killed off the susceptible bacteria, effectively breeding generations of superbugs that are resistant to our antibiotics.

On the other hand, we shouldn't be too hard on ourselves. After all, the goal of a microbe in producing and using its own antibiotic against its neighbors is very different than the goal of physicians in prescribing antibiotics. A single bacterium produces and uses just enough antibiotic to allow it to thrive against its competitors. Theirs is a selfish use of antibiotics. In contrast, physicians prescribe antibiotics to help our patients. Ours is a selfless use of antibiotics, which by its very nature *must* lead to a much broader environmental exposure to antibiotics than a single bacterium producing antibiotics only for its own benefit. There are also issues of scale. Secretion into the environment of a trillionth of a gram of antibiotic may well be enough to protect a single, tiny bacterium from its neighbors. But a trillionth of a gram of antibiotics would do nothing for a patient. Rather, every day of antibiotic therapy requires that a billion or a trillion times more antibiotic (i.e., milligram to gram quantities) be administered to cure infections in patients.

Of course, inappropriate antibiotic use—for example, prescription of antibiotics for viral rather than bacterial infections—contributes to the spread of drug resistance but does not help the patient either. Inappropriate antibiotic use must be curtailed to help us slow the spread of antibiotic resistance, which will buy us more time to come up with a long-term solution to the problem.

The melanic and peppered moth example of natural selection is also of interest because it suggests that it might in fact be possible to cause the percentage of antibiotic-resistant microbes to go in the other direction (i.e., lower) if we could curb our antibiotic use. After all, the British caused a reduction in the numbers of melanic moths, and an increase in the numbers of peppered moths, by reducing their release of pollutants into the air. This is, indeed, the logical basis for those who argue that we are responsible for antibiotic resistance, and that we could conquer resistance if we tried hard enough.

But there are several problems with the presumption that more-appropriate use of antibiotics is going to allow us to finally defeat microbes.[23] First, there is a major, practical difference between the process of Industrial Melanism in moths and the process of antibiotic resistance in bacteria. Industrial Melanism was caused by undesirable pollutants. Antibiotic resistance is driven by use of antibiotics, which are not undesirable pollutants. Yes, it may be possible to reduce the amount of antibiotics society uses, but a large component of antibiotic use is not only appropriate, it is necessary. We simply cannot reduce antibiotic use in infected, sick patients. As the global population continues to age, and as critical care and other high-tech medical advances are made, there are going to be more and more people who get bacterial infections, which are going to require more and more appropriate antibiotic use. So, there is a floor of antibiotic use below which it is not desirable for us to go. In contrast, it is desirable to reduce pollutants to as low a level as possible, since they have no useful function. The British were able to effect a massive reduction in airborne pollutants, and reduction of antibiotic use on that same scale is likely not feasible.

A second problem with the suggestion that Industrial Melanism demonstrates antibiotic resistance can be conquered by reducing our antibiotic usage is more fundamental. Note that melanic moths have not been eliminated from the population of moths, despite massive reductions in pollution levels. Melanic moths are still found at perhaps 5 percent of the overall moth population. They are reduced but not gone, and the moment that pollution comes back, you can be certain that melanic moths will once again take over. Melanic moths are capable of persisting despite elimination of relevant pollutants because pollutants did not *cause* the melanic moths to become melanic. Melanic moths existed long before humans caused pollution problems. Pollution only changed the frequency of melanic moths in the population, but did not actually cause the moths to exist in the first place.

Similarly, our use of antibiotics does not actually *cause* the resistance to occur. Rather, our use of antibiotics affects the frequency with which preexisting resistance mechanisms are spread through bacterial communities by killing off susceptible bacteria, leaving the resistant ones behind to reproduce. It is well known that once antibiotic-resistance genes are established at a certain level within a population of bacteria, those genes remained fixed in the population and will become resurgent whenever the bacteria are again exposed to the antibiotic.[24]

This distinction between our causality of microbial resistance and our affecting the rate of spread of resistance must be recognized if we are to create a true solution to the problem of antibiotic resistance. If our misuse of antibiotics causes drug resistance, the solution that would allow us to forever defeat microbial resistance would be for us to strictly use antibiotics only when truly indicated. On the other hand, if our misuse of antibiotics affects the rate of spread of resistance, but does not actually cause resistance, then using antibiotics correctly will not stop microbial resistance—it will only slow it down so that we can find a real solution to the problem.

Framed in this context, it is clear that convincing physicians to use antibiotics properly is an important step to take, not because it is a permanent solution to drug resistance but because it will buy us more time to create a real solution to the problem. As Dr. Scott Gottlieb, the former deputy commissioner of the FDA, has written, "Preventative efforts alone won't solve our bacterial challenges. What we need most [is] . . . new medicines."[25] That is, the only viable, long-term solution to the problem of microbial resistance is to concede that we will never truly defeat it, that we can only keep pace with it, and that to do so, we must have a continuing, steady development of new antibiotics in perpetuity. These concepts have been summarized succinctly and precisely by Nobel laureate Dr. Joshua Lederberg: "The future of humanity and microbes will likely evolve as . . . episodes of our wits versus their genes."[26]

So, clearly society is in need of a steady stream of new antibiotics to allow us to keep pace with microbial resistance. The question is, why aren't we getting new antibiotics anymore?

WHY AREN'T ANTIBIOTICS BEING DEVELOPED ANYMORE?

In the words of Dr. C. Hal Jones, a leading scientist who worked on antibiotic development at the pharmaceutical giant Wyeth: "Keeping up with evolution is a tough business."[27] Dr. Jones made this statement in a lecture at the IDSA's annual meeting in 2007. Dr. Jones was referring to the need for society to keep pace with the evolution of microbial resistance by developing new antibiotics. The problem is—as Dr. Jones so eloquently expressed—it isn't easy to keep up with the microbes. It would be hard enough to keep up with microbes if we were trying. But the amazing thing is, we're not even trying anymore. Ironically, a year after Dr. Jones gave his talk in which he described the difficulties of conducting antibiotic research in the pharmaceutical industry, Wyeth shut down its entire antibiotic development program and Dr. Jones's research along with it.

The cause of the decline in antibiotic development is multifactorial, but fundamentally each factor relates to return on investment—that is, how much money is made on sales of a drug relative to the amount of money invested in discovering and developing it.[28] Unfortunately, in the twenty-first century, drug development in general is facing increasing challenges, and these challenges are magnified for antibiotics.

The first barrier to developing a new drug is the incredible scientific complexity involved in identifying lead candidate molecules. In the 1990s, it was thought that cutting-edge molecular-biological techniques were going to enable far more efficient drug discoveries by allowing companies to use rational, targeted, "designer drug" strategies. Unfortunately, cutting-edge scientific techniques have not sped up or made less expensive the development of new drugs in general, or new antibiotics in particular.[29] If anything, in the face of the molecular revolution, fewer new drugs are being developed in all classes—not just antibiotics—as compared to previously,[30] at least in part because the cost of new drug development has skyrocketed, and has not declined as predicted.[31] Even the US Government Accountability Office and the US Food and Drug Administration have independently come to this conclusion.[32]

A major exception to the declining efficiency of new anti-infective development is the development of new medications to treat HIV. As I will discuss in more depth a little later, there have been numerous breakthrough medications for the treatment of HIV that were developed using cutting-edge molecular technology. But, alas, not so for new antibiotics to treat bacterial infections.[33]

After investing billions of dollars in high-tech molecular biology research labs in the 1980s and 1990s, those few pharmaceutical companies still developing anti-infectives have largely gone back to the tried-and-true method by which they used to develop antibiotics: mass screening.[34] These companies own literally millions of chemical molecules. They take those molecules and test them in mass screening studies to see if any of them kill microbes growing in the laboratory. It's a simple, brute-force approach. Since we know that antibiotics were actually invented by microbes, and our primary role is to plagiarize the ideas microbes have already come up with, another strategy is to harvest microbes from soil or other sources, and expose them or their by-products to recipient microbes, to see if anything is found that kills the recipient microbes. Neither of these strategies has much to do with rational, targeted drug discovery.

Let me offer a specific example of a publicly reported failure of rational, targeted antibiotic discovery using cutting-edge molecular genetics. In 1995, the complete genome (the entire genetic sequence of all of an organism's genes) of the bacterium *Hemophilus influenzae* (a common cause of pneumonia and other types of infections) was published. Of course, with publication came the expectation that miracle cures would derive from the knowledge of the entire genetic sequence of that bacterium. So, from 1995–2001 GlaxoSmithKline, one of the largest pharmaceutical companies in the world, undertook a high-tech, molecular-genetics study of more than three hundred potential bacterial gene targets against which antibiotics might be developed.[35] GlaxoSmithKline initiated no less than sixty-seven separate, intensive development campaigns against those three hundred gene targets. Of the sixty-seven research and development campaigns initiated—each costing approximately one million dollars—only sixteen campaigns identified a gene thought to be of possible relevance to developing an antibiotic. Unfortunately, of those sixteen gene targets identified in initial laboratory screens, only five panned out to be real. And of those five targets, no drugs could be designed to inhibit them. So, after a seven-year, multimillion-dollar effort, nothing was gained.

Subsequently GlaxoSmithKline's antibiotic efforts returned to old-fashioned screening of their massive chemical library. Over the subsequent five years, this brute-force, inelegant, old-school screening identified six drugs that may be developed in the future.[36] This scenario, publicly described by a GlaxoSmithKline scientist, underscores not only the difficulty in drug development but also the lack of the ability of cutting-edge technologies to improve the process. Indeed, other companies have had the exact same problems trying to use high-tech molecular genetics to identify new antibiotics.[37]

In retrospect, it was rather naive to believe that merely knowing the genetic sequence of an organism was immediately going to allow new therapies to be developed to treat infections caused by that organism. The genetic sequence of an organism is composed of a DNA strand in which the DNA nucleotides adenine (A), thymine (T), cytosine (C), and guanine (G) are strung together in a specific sequence. The sequence of the DNA nucleotides creates the overall "genetic code" of an organism. That genetic code of DNA nucleotides is translated inside the organism into amino acid sequences that are then strung together to form proteins. Thus, it is the organism's DNA sequence that encodes all of the proteins that the organism can make. Therefore, knowing the DNA sequence (e.g., ATTACGACGT . . .) of a bacterium's entire genome allows us to know the sequences of amino acids that comprise all the proteins the bacterium can produce. But how does knowing the amino acid sequence of a bunch of proteins allow us to develop drugs that are going to kill a bacterium?

Just because we know the amino acid sequence of a protein doesn't mean we know the shape or function of the protein. Nor does it mean we will know how to design a drug to inhibit that protein's function. Nor does it mean we will know that inhibiting that protein will cause the bacterium producing it to die. Nor does it mean we will know that such a drug, even if it did successfully inhibit that protein and thereby kill that bacterium, will be nontoxic to people. As you can see, there are multiple critical drug-development layers that are required to be completed after one has figured out the amino acid sequence of bacterial proteins. And, of course, for the purposes of some semblance of simplicity, I've left out a number of even more complicated scientific barriers to drug development.

Knowing the genome sequence of a microbe (or of a person, for that matter) is important from a basic science perspective. It allows all kinds of interesting experiments to be performed, and it lays the groundwork for discovery of treatments for diseases at some point in the future. But it offers no immediate insight into how to create treatments for diseases. I would say that knowing the genome sequence of an organism is akin to discovering one portion of the Rosetta stone, the portion with the Egyptian hieroglyphics on it. So, you have a long string of hieroglyphics that are written as if to actually say something intelligent. But the problem is, you don't know what it actually says. And you need to spend years trying to figure it out because you don't have the Greek translation on the other portion of the stone.

A critical question is, why has a targeted, basic science investment in understanding the biology of HIV enabled such an enormously successful drug-development portfolio aimed at this virus, while there has been no similar success in developing new antibacterial agents? If we could understand this difference between how we've studied HIV and bacteria, it should shed light on what we should do differently to develop new antibiotics targeting bacteria. As we shall see, the primary differences explaining a lack of similar success in developing new antibiotics and new anti-HIV drugs are: (1) an insufficient focus of funding and manpower on basic research into antibiotic-resistant bacteria; (2) the absence of focus on "translational research" in antibiotics (that is, research intended to "translate" basic science discoveries about how bacteria grow and live into practical drugs that can be used in patients to treat bacterial infections); and (3) the absence of a federally funded, comprehensive network of investigators that would enable large and expensive clinical trials for new antibiotics to be conducted in an efficient manner.

So, for now, companies have had to go back to slogging through enormous libraries of molecules or microbes, in the hope of random discoveries of new efficacy from those molecules.

Ultimately, discovering an antibiotic compound is only the first step in the eventual development of a new drug. Not only does a discovered compound have to kill microbes in the test tube, but the compound also has to be safe and effective in animals. And it also has to be "drug-able." "Drug-able" means, for example, that the compound has to be active at concentrations that are feasible for administration to people. It has to be absorbable into the body. And it has to be practical to manufacture and prepare the drug for a reasonable cost. For instance, a recent antifungal compound was effective at killing fungi in the test tube. The problem was that to achieve the concentration necessary to kill fungi in the body, a person would have had to ingest literally kilograms (a kilogram = 2.2 pounds) of that compound. The compound worked, but it was just not "drug-able." Its development has been terminated.

The next barrier to developing a new drug is proving that the compound is safe and effective in human beings. In order to meet FDA requirements for establishing safety and efficacy, a compound must be tested in three phases of human clinical trials: phase I (safety), phase II (a study to figure out what doses of the drug are safe and possibly effective), and phase III (the large, definitive study to prove that the drug is effective). Each of these trials is expensive and takes considerable time and effort to complete. In particular, phase III clinical trials are outrageously expensive. A recent analysis found that the average cost of a phase III clinical trial was $86 million ($105 million in 2008 inflation adjusted dollars).[38] Even the earlier phase I and phase II trials can cost in the millions to tens of millions of dollars.[39] Given the incredible expense of phase III trials, pharmaceutical companies are understandably conservative about selecting which drugs to take into such a trial. Furthermore, the prohibitive cost of these trials makes it very difficult for smaller companies or academic researchers to carry them out.

The overall success rate of developing a new drug is only 1 in 10,000 molecules initially tested in the laboratory as a drug candidate.[40] Even those molecules that progress through the entire preclinical development phase (i.e., testing in the laboratory, testing in animals, proper manufacturing, etc.) and that are ultimately tested in clinical trials are highly likely to never make it through the clinical testing and approval process. Specifically, a drug that is safe and effective in animals and is drug-able still has about a 90 percent chance of failing during either phase I, phase

II, or phase III clinical trials.[41] Drugs can fail during clinical trials because they are not safe, because they are not effective, or both. Drugs can even fail because they are effective, but their efficacy isn't large enough to justify a corporate investment of hundreds of millions of dollars for further development. As you can see, it's a helluva thing to develop a drug. Given all of the above costs and barriers, and the high chances of failure of multiple drug candidates, it costs well over one billion dollars (in 2008 inflation adjusted dollars) in research and development and lost opportunity costs for each drug that is ultimately found to be safe and effective, and is approved by the FDA for use in humans.[42]

Thus far, we have discussed barriers to drug development in general, and they are clearly considerable. But despite these barriers, pharmaceutical companies are still developing drugs. Drugs targeting cancer, arthritis and inflammatory diseases, cardiovascular diseases (like high blood pressure, high cholesterol, etc.), diabetes and other endocrine diseases, chronic lung diseases (like asthma or Chronic Obstructive Pulmonary Disease from smoking), and other diseases of aging (such as pain, dementia, etc.) are being developed at a reasonable rate.[43] So, what makes antibiotics different?

Unfortunately, antibiotics have a lower relative rate of return on investment compared to other drugs.[44] Why? A major reason is that antibiotics are short-course therapies that cure their target disease, and hence are typically taken for no more than one to two weeks. In contrast, chronic diseases are treated with noncurative therapies that suppress symptoms and are required to be taken for the life of the patient. For example, patients have to take cholesterol-lowering drugs or drugs for high blood pressure for the rest of their lives. Ironically, antibiotics are victims of their own success: they are less desirable to drug companies and venture capitalists as targets for development because they are more successful (that is, effective) than other drugs.

As noted, a dramatic illustration of the power of chronicity of therapy in driving interest in drug development is the remarkable and continuing success of the development of new therapeutics to treat HIV infection. It is worth spending a few moments reviewing the success of developing drugs to treat HIV, as an example of what scientists, technologists, and clinical-trial investigators can achieve when each are properly focused and funded.

Antiretrovirals used to treat HIV are excellent examples of therapeutic agents that are taken chronically for the remainder of a patient's life, and are typically initiated in relatively young patients. The combination of a lifelong treatment that is initiated at a young age results in big profits for companies that make these drugs. Therefore, technologists are interested in developing these products, and over the last fifteen years, pharmaceutical companies have brought to market virtually the same number of new drugs targeting just HIV as for the treatment of all bacterial infections combined. In particular, over the last five years considerably more new HIV drugs than antibiotics have been developed.

Just as compelling is the number of new anti-HIV medications we found to be in development by the largest pharmaceutical and biotechnology companies in 2004.[45] At that time, there were more than twice as many drugs in development to treat just one virus, HIV, as compared to all bacterial infections combined. That trend is continuing five years later.

A review of the medications used to treat HIV underscores the importance of laying a fundamental bedrock of basic science knowledge to enable platforms of technology to be developed. Over the last twenty-two years, there have been no fewer than twenty-five medications specifically developed to treat HIV infection.[46] These medications are divided into at least six different classes, with members of each class working to poison different biochemical pathways used by HIV to copy itself and/or infect human cells.

Some anti-HIV drugs, called nucleoside reverse transcription inhibitors (NRTIs), are designed to poison the ability of HIV to copy its own DNA; these drugs are specifically custom designed to poison the HIV's DNA-copying mechanism while sparing the DNA-copying mechanism used by human cells. Examples of these drugs include the famous AZT, which was the first drug brought to market to treat HIV back in 1987.[47]

Other anti-HIV drugs, called non-nucleoside reverse transcription inhibitors (NNRTIs), are also designed to poison HIV's DNA-copying mechanism, but they work by blocking a completely different protein site than NRTIs. The importance of blocking a different protein site is that when HIV becomes resistant to NRTIs, it does not automatically become resistant to NNRTIs, and vice versa. This means you can combine NRTIs and NNRTIs to enhance each other's effects, and can use NRTIs to treat NNRTI-resistant viruses, and/or use NNRTIs to treat NRTI-resistant viruses.

Another class of HIV medications, called protease inhibitors (PIs), has been designed to block a protein that HIV uses to construct its infectious form. So PIs block a part of the viral life cycle that is completely unrelated to NRTIs or NNRTIs. It was, in fact, the introduction of the first protease inhibitors in 1995 (saquinavir, ritonavir, and then indinavir) that really revolutionized HIV treatment. By combining PIs with NRTIs, physicians for the first time were able to prevent the virus from copying itself for prolonged periods of time. By suppressing the virus from copying itself, the drugs enabled the infected patients' immune systems to rebuild themselves. Combination anti-HIV drug regimens that used multiple drugs from different classes to treat the infection came to be called Highly Active Anti-Retroviral Therapy (HAART). Because of the incredible success of HAART, we now consider HIV to be a chronic, treatable infection, like diabetes, rather than a uniformly fatal infection as it was previously considered.

Since the advent of PIs, several other completely independent types of antiretrovirals have been developed. For example, enfuvirtide blocks the ability of HIV to force its way inside human cells by inhibiting the interaction between a viral protein and human cell membranes. Maraviroc is another drug that also blocks HIV entry into the cell, but does so by blocking the interaction between HIV and a completely different protein (called "CCR5") than enfuvirtide. Raltegravir works in a totally different site altogether, having nothing to do with blocking HIV entry into cells. Rather raltegravir blocks the ability of HIV to insert itself into human DNA and stably infect cells.

As these brand new drugs, with completely new mechanisms of action, have come onto the market, drug companies have also gone back and developed newer versions of old classes of drugs that work better than the original drugs. For example, etravirine is a new version of NNRTIs. The importance of etravirine is that it is effective against HIV, which has become resistant to all prior NNRTIs. Similarly, new PIs called darunavir and tipranavir are also active against HIV that is resistant to prior PIs.

The antiviral armamentarium now available to treat HIV infection is nothing short of awesome. It is astounding in its efficacy. It is amazing in its safety, tolerability, and convenience. And it is awesome in that the full spectrum of drugs now available can allow at least some treatment for virtually any drug-resistant virus. People who die from HIV nowadays are those who are unable or unwilling to take these effective medications on a regular basis, or those who develop cancer or heart disease, or other chronic illnesses.

In stark contrast to antibiotics, the incredible diversity and power of anti-HIV drugs owes itself almost entirely to: (1) a large investment in basic science to understand how HIV causes disease; (2) corporate economic motivation to sell chronic drug therapy for HIV infection; and (3) an enormously productive and successful large network of NIH-funded clinical investigators whose mission is to conduct clinical trials and other translational research into HIV infection.

None of the many anti-HIV treatments would have become available had billions of dollars not been spent to conduct basic scientific research into the biology of HIV. We had to understand how HIV replicates itself, how HIV gets into cells to cause infection, how it inserts its own genes into human DNA, how the virus manufactures itself inside cells, and how it leaves cells to cause infection in new cells. We had to know exactly how HIV binds to and penetrates inside of human cells. For us to be able to design specific molecules to block the virus from binding to and getting inside of our cells, we had to know exactly what proteins on human cells are used by HIV to enter into our cells. We also had to know the exact three-dimensional shapes of those proteins so we could custom-design molecules to block the interaction between HIV and those proteins.

The scientific information describing all of the above is what enabled pharmaceutical companies to develop specific technologies to poison each step in the virus's life cycle. The motivation to sell effective drugs that patients take for the rest of their lives induced companies to spend their capital in the development of those drugs. Finally, the AIDS Clinical Trial Group (ACTG) network has played an integral role in the successful design, execution, and analysis of clinical trials for virtually all of the anti-HIV drugs that have been developed. The ACTG is a network of clinical-trial investigators at academic research centers all over the country and indeed throughout the world, which is funded by federal grant dollars from the National Institute of Allergy and Infectious Diseases (NIAID) at the NIH.[48] The ACTG was formed in 1987, and since then it has run and continues to manage a dizzying array of highly successful, cutting-edge translational and clinical research into disease caused by and new treatments for HIV.[49] The availability of the ACTG enables clinical investigators at research centers throughout the world to participate in clinical trials and other important translational HIV research. Without the ACTG, the volume, success rate, and efficiency of such research would be far less, and it is likely that fewer anti-HIV drugs would have been developed.

With the basic science research, economic motivation, and available clinical trial networks, the time it has taken the biomedical scientific community and pharmaceutical industry to achieve creation of our anti-HIV armamentarium is nothing less than remarkable. The first cases of what would eventually be called AIDS arose in the United States in 1981 and 1982. The virus causing AIDS was first isolated in early 1983 by Dr. Luc Montagnier and his colleagues, working at the Pasteur Institute in France.[50] By 1984, Montagnier's discovery had been confirmed by Dr. Robert Gallo[51] and Dr. Jay Levy,[52] working in the United States. Shortly thereafter all of the genes in HIV were sequenced. Within four years, the first effective drug, AZT, was developed. As Dr. Montagnier has written, the virus causing AIDS was identified within two and a half years of the first presentation of patients with the disease, within two more years a blood test was developed to diagnose and detect the virus causing the infection, and within another two years the first treatment for the disease became available.[53] The early history of the medical treatment of HIV is nothing less than an astonishing tribute to the success of focused, high-quality basic science, in partnership with highly motivated private corporations and federally funded clinical trial networks that translated the basic science into useful technology.

Within eleven years of confirmation of the virus causing AIDS, a complete revolution in treatment became available (i.e., HAART). During the subsequent decade, additional breakthrough treatments were derived. Original HAART treatment required taking a dozen or more pills multiple times per day, with significant toxicities. The pills were difficult to tolerate. Now, newer versions of those drugs are far less toxic. They are easier to tolerate. They can be taken twice, or now even once per day. Furthermore, there are combination pills on the market, with multiple medications in each pill. It is actually now possible to treat HIV infection by having patients take one pill once per day that has few side effects. These treatments are nothing less than an absolute revolution in the care of patients infected with HIV. Indeed, in many ways it is now easier to treat HIV than diabetes. Like diabetes, HIV remains a chronic illness (i.e., we can't cure HIV, but we can control it with chronic medications). But with twenty-first-century medications, patients infected with HIV should live life spans that are quite similar to those of HIV-uninfected patients.

Collectively, these facts are critical when considering strategies to stimulate antibiotic development, because they clearly demonstrate, understandably, that when financial advantage is apparent to pharmaceutical companies, they are still capable of and interested in making anti-infective agents. Furthermore, they demonstrate that a massive, focused basic science effort can lay a critical foundation for future drug discovery that enables miraculously effective anti-infective drugs to be developed. Finally, they illustrate the importance of making funding available to clinical investigators who are capable of designing and executing clinical trials and related translational research to study new treatment strategies for infections.

To its credit, the NIAID at NIH is now emphasizing funding for research on antibiotic-resistant bacteria on a far greater scale than previously.[54] However, the basic science and translational research funding for antibiotic-resistant bacteria is still far less than for HIV infection. Furthermore, there is little funding going to clinical trials for antibiotics, and there is still no network of clinical trial investigators to conduct clinical trials of treatments for drug-resistant bacterial infections, akin to the ACTG for HIV treatment. In an era where MRSA, by itself, kills more Americans every year than does HIV, not to mention the even more deaths being caused every year by other drug-resistant bacteria, it is increasingly difficult to understand why there is not an NIH-funded network of investigators to study antibiotic clinical trials. Such a network is desperately needed for the development of treatment strategies for antibiotic-resistant bacterial infections.

As Dr. Lou Rice, chairman of the IDSA's Research on Resistance Work Group, has noted, of the $500 million or so that the NIAID/NIH spends on drug-resistant infections, only approximately 10 percent is spent on bacterial infections, with the rest being spent predominantly on HIV research. Dr. Rice wrote, "In each of these areas (optimal antimicrobial therapy, infection control, physician and patient behavior), our evidence-based knowledge in the area of HIV (accumulated for < 3 decades) far outstrips our knowledge in the area of antibacterial therapy (around for nearly 8 decades). . . . To truly succeed, [we] need . . . a much larger coordinated effort [on antibacterial research], one with considerably more resources devoted to it than at present."[55]

Aside from economic considerations that make antibiotics undesirable, an additional major problem with discovering new antibiotics relative to other types of drugs is explained by the "low-hanging fruit" theory. There have been well over one hundred antibiotics discovered and developed for use in humans in the United States over the last half-century, and more than 150 antibiotics throughout the world. That means many of the "easy"-to-develop antibiotics have already been identified and developed. Each subsequent generation of antibiotics becomes harder and harder to develop, akin to picking the fruit lowest on the tree, leaving behind the harder-to-reach fruit on the upper branches.

Another factor that weighs heavily as a disincentive for antibiotic development is the appropriate public-health need to limit use of new, broad-spectrum antibiotics. Antibiotics are absolutely unique among all classes of drugs in that they become less effective the more they are used. Yes, it is true that chemotherapy agents may become less effective after multiple cycles of treatment in individual patients, as a patient's cancer starts to become resistant. But one patient's cancer does not transfer its resistance to the next patient, so the chemotherapy demonstrates no loss of activity in the next patient. In contrast, antibiotic resistance is transmissible from person to person.

I am not aware of any other technology on Earth that similarly becomes less effective in a transmissible manner the more it is used (the closest example is pesticide resistance, but this is not transmissible from patient to patient). For example, this would be the equivalent of your computer working less effectively each day after you buy it. Not only that, every time your neighbors used their computers, your computer would also become less effective. And your computer's loss of activity would be directly transferred to my newly purchased computer, and would make my new computer work less effectively. In fact, the more people using computers, the faster all of the computers would become less effective.

Given the transferable loss of activity of antibiotics the more they are used, it is understandable why leading infectious-diseases specialists strongly encourage restriction of the use of newly released, powerful antibiotics. Indeed infectious-diseases specialists perceive our stewardship of antibiotic use as being a primary role of our specialty.[56] But, we must acknowledge that there is an extremely important and highly undesirable side effect of our appropriate antibiotic stewardship. When we discourage use of new antibiotics in order to prolong their useful lives, we inevitably negatively impact sales of new antibiotics.[57] In direct contrast, when new drugs in other classes become available, their use may be encouraged by medical leaders, resulting in enhanced sales. Pharmaceutical companies obviously know this, and it is yet another reason why antibiotics are less desirable for development compared to other classes of drugs.

Finally, an issue that is repeatedly cited by both pharmaceutical and biotechnology companies as a major deterrent for the development of antibiotics is the lack of available guidance documents from the FDA as to what type of clinical trials the agency considers acceptable to demonstrate the safety and efficacy of new antibiotics.[58] Since companies don't know exactly what the FDA is looking for in clinical trials, and what the FDA considers acceptable seems to change from year to year, companies are taking a gamble when they design antibiotic clinical trials. They run the risk of spending three years and $100 million on phase III clinical trials in which safety and efficacy of the new antibiotic is proven, only to have the FDA come back and say, "We don't like how you designed this study, so we are not going to accept these data and we are not going to approve your drug for use in humans."

Believe it or not, this scenario actually happens. A recent example occurred when the biotechnology company Oscient asked the FDA to approve the use of its antibiotic, gemifloxacin, for the treatment of sinus infections. Numerous antibiotics had previously been approved for the treatment of bacterial infections of the sinuses. So, Oscient wasn't exactly blazing a new trail. To the contrary, they were following an exceedingly well-worn path, which had previously rarely failed to lead a new antibiotic to the promised land of receiving approval for use in humans. Oscient simply carried out clinical trials similar to all the previously conducted clinical trials for sinusitis treatment, and showed their antibiotic worked as well as other antibiotics that had been previously approved to treat sinus infections.

Oscient submitted the results of the successful trials to the FDA for review. Unfortunately, just as Oscient was submitting their application, the FDA had begun rethinking its policy on accepting data for clinical trials of many infections, including sinus infections. As a result, Oscient and their clinical trials got slammed, and it wasn't their fault. Their clinical trials had been completely acceptable when they were originally designed. The trials were completely acceptable even when the company began enrolling patients into them. But, suddenly what would have been deemed acceptable a year earlier was no longer sufficient to get an FDA indication, just as Oscient finished their trials and submitted their results to the FDA.

How big a blow was this? Oscient had spent years and tens of millions of dollars conducting no fewer than five separate clinical trials into which 1,800 patients with sinus infections had been enrolled to prove their antibiotic worked for sinusitis treatment. Rejection of their application was a tremendous blow, which negatively impacted Oscient's stock price and its capitalization. Furthermore, the impact went far beyond slamming Oscient. The decision to reject Oscient's application for sinusitis sent a tremor through the entire antibiotic-development world. Oscient had no way to know, when they started their trials, that eventually the FDA was going to change the way it viewed data necessary to prove efficacy against sinusitis. You can imagine that other companies watching this all unfold began wondering what the FDA would do to the standards for proof of efficacy of antibiotics for the treatment of other diseases. Those companies likely began rethinking plans to spend more money developing new antibiotics to treat those other diseases.

Why did the FDA reject Oscient's application to approve their new antibiotic for the treatment of sinusitis? Because the FDA is under increasing pressure from political forces to crack down on pharmaceutical companies. As the experts began to review the available literature on the treatment of sinusitis, it began to be realized that sinusitis often simply resolves on its own after a while. The question became, is any antibiotic truly effective at treating sinusitis, or does sinusitis always just get better on its own anyway? If sinusitis always just gets better anyway, an antibiotic would look great in the trial, because all the patients would get better even though it wasn't the antibiotic making them get better. How could a company prove that their antibiotic was what was making patients get better from sinusitis? In essence, the FDA was saying, "We know that you can prove a new antibiotic is as effective as older antibiotics at treating sinusitis. But how do we know that the older antibiotics were more effective than no treatment? And if we don't know that, how do we know that a new antibiotic is more effective than no treatment?"

The way to answer this question would be to compare the effect of an antibiotic to a placebo (i.e., a pill that looks like the antibiotic but in fact has no medicine in it) for the treatment of sinusitis. If patients receiving antibiotics did not do better than patients receiving the placebo, you could prove that antibiotics do not work for sinusitis. If patients receiving antibiotics did do better than patients receiving the placebo, you could prove that antibiotics do work for sinusitis. So the FDA decided that since it did not have evidence that antibiotics were more effective than placebo to treat sinusitis, it could no longer approve new antibiotics that were shown to be as effective as old antibiotics to treat the disease.

As it turns out, there already are studies in the medical literature that have compared the effect of antibiotics to placebo in the treatment of sinusitis. Indeed, a recently published analysis of the effect of antibiotics versus placebo for the treatment of sinusitis actually found that antibiotics do make sinus infections go away faster.[59] The authors analyzed multiple previous trials that had compared antibiotics to placebo. Each study by itself had not been able to confirm definitively that antibiotics worked better than placebo. But each study had been small, too small to have had enough patients in them to prove an effect of antibiotics. So the authors of this new study lumped the data together from all those previous, individual studies. When the authors analyzed the data from all the prior studies lumped together, they found that antibiotics increased by 35 percent the probability of curing sinusitis in one to two weeks. These authors unequivocally found that antibiotics were superior to placebo in causing sinusitis to get better faster.

Yet the authors then went on to create a computer analysis of 10,000 theoretical patients with sinusitis. That is, they asked the computer to predict what would happen to theoretical patients with sinusitis who were treated with theoretical antibiotics. Their computer analysis of these theoretical patients who didn't actually exist suggested that they could not be certain that antibiotics were effective for sinusitis. So, the authors seemed to ignore their own findings that resulted from analyzing actual clinical trials of real patients with sinusitis treated with real antibiotics, and in their conclusion instead focused on the negative finding from their computer model of hypothetical patients. They suggested that antibiotics shouldn't be prescribed for sinusitis because the computer model couldn't confirm that antibiotics worked, even though the patient data showed that antibiotics did work!

My conclusion from the study is that antibiotics work for sinusitis, because I trust data based on use of real antibiotics to treat real patients more than I trust computer models of hypothetical antibiotics used to treat virtual patients. There's no getting around the fact that when you add together the results of nine prior trials in which antibiotics were compared to placebo for treating sinusitis, antibiotics increased by 35 percent the probability that patients with sinusitis would be cured within one to two weeks of starting the treatment.

The question of whether or not antibiotics work enough to justify their use to treat sinusitis is a separate question. Maybe it doesn't matter that sinusitis resolved more quickly with antibiotics than without. One could argue that society has a moral imperative to restrict antibiotic use to save these drugs for people with life-threatening infections, and that we shouldn't be wasting them to treat infections that, while uncomfortable, only rarely cause dangerous effects and almost always eventually get better on their own even without treatment. Perhaps that argument is correct. On the other hand, that has not been the conversation that the medical community has undertaken to date. Perhaps it is time to rethink our stance on whether antibiotics should receive an FDA indication to treat sinus infections, taking into consideration a known efficacy of the drugs for sinusitis balanced against the risk of inducing antibiotic resistance by using antibiotics to treat nonlethal infections.

Of course, any reconsideration would come too late for Oscient's gemifloxacin. And the message sent to industry—that regulatory standards for antibiotics are subject to change at any time—has already caused damage to future development of all antibiotics, including those that might be used to treat life-threatening infections.

Unfortunately, the rethinking of FDA standards for antibiotics did not stop at sinusitis, which doesn't kill people. The rethinking of antibiotic standards subsequently progressed to include pneumonia (i.e., lung infection), which does kill people. Believe it or not, in 2007 there was a strong movement afoot to say that new antibiotics could not be approved to treat pneumonia unless they were shown to be better than placebo. Think about this. Can you imagine taking a patient with pneumonia and asking him or her to participate in a research study in which he or she could be treated with placebo instead of an antibiotic? Pneumonia kills people! How would you like to have pneumonia and receive treatment with placebo? How would you like your child with pneumonia to receive treatment with placebo? How could any physician, in good conscience, consider administering placebo to a patient with pneumonia?

This rethinking of standards for approval for new antibiotics to treat pneumonia was an extremely dangerous development. Aside from the astounding ethical implications of a government regulatory agency insisting that physician-scientists administer placebo to patients with potentially life-threatening infections, had the FDA insisted that placebo-controlled studies of pneumonia be conducted, it would have threatened a major cornerstone market for new antibiotics. No pharmaceutical company would have been willing to conduct such studies. They simply would have stopped developing new drugs to treat pneumonia. And if we lost the pneumonia market, it would have been a near deathblow to the future of antibiotic development in general.

As bizarre as it seems, this rethinking of antibiotic standards for pneumonia was really not the fault of the FDA. The FDA is in an impossible situation. They are like the offensive line of a football team. You only notice them when something goes wrong, and you don't notice them the 99 percent of the time when things go right. It is that rare instance when things go wrong that politicians and the media jump all over the FDA. The people working at the FDA get flack from all sides, including the drug companies, practicing physicians, politicians, and the press. The people at the FDA are doing their best, day in and day out, to make sure that drugs are safe and effective. But clinical medicine is so complicated nowadays that it can be very challenging to establish criteria that will ensure that drugs are safe but still enable companies to develop new drugs.

The situation that the FDA found itself in with respect to antibiotics to treat pneumonia was an impossible one. Could they really insist that new studies for pneumonia be designed to prove that antibiotics were better than placebo? It was at this point that Dr. Ed Cox, director of the Office of Antimicrobial Products at the FDA, agreed to hold a joint workshop with the IDSA to explore the issue of how new drugs for the treatment of pneumonia could be approved. That is, to their credit, the FDA went to the national infectious-diseases experts and asked for their expertise and input to help resolve this question.

On January 17 and 18, 2008, the FDA and IDSA cosponsored the "Clinical Trial Design for Community-Acquired Pneumonia" workshop.[60] The workshop provided an opportunity for experts from the United States and Canada to discuss how new antibiotics could be proven to be effective for the treatment of pneumonia, and therefore how we could enable pharmaceutical companies to continue to develop such badly needed antibiotics.[61]

One of the fundamental questions discussed at the workshop was, is it ethical to enroll patients with pneumonia into a clinical trial in which they might get treated with placebo rather than an active antibiotic? The nearly uniform consensus among the physicians present was, no, it was not ethical. In fact, many physicians were flabbergasted that such a thing could even be contemplated. Nevertheless, nagging questions remained. The real issue was, had anyone already proven that antibiotics were more effective than placebo for the treatment of pneumonia? The prevailing opinion before the meeting was that no one had ever compared the effect of antibiotics for pneumonia to placebo/no treatment, because it was unethical to do such trials. Nobody at the workshop, including senior and internationally renowned infectious diseases specialists, could recall such trials having been done.

But, in fact, as a result of research conducted in anticipation of the coming workshop, Dr. Mary Singer of the FDA discovered that indeed such trials had been done at the beginning of the antibiotic era. I've described to you in earlier chapters how effective antibiotics are at treating infection. It should come as no surprise that the physicians who were taking care of patients in the 1930s and 1940s understood exactly what they were dealing with when antibiotics came along. They weren't dummies. They conducted clinical trials that showed that antibiotics worked.

When FDA personnel and IDSA members began to dig into the historical literature from the 1930s and 1940s, we found numerous absolutely seminal, classic studies that time had unfortunately forgotten.[62] When we summed up all those studies, we found that antibiotics massively reduced the risk of dying from pneumonia compared to no treatment.[63] Specifically, for patients aged 12 to 29 years, the risk of dying from pneumonia fell from 12 percent with no treatment to 1 percent with antibiotic treatment. For patients between the ages of 30 and 59 years, the risk of dying from pneumonia fell from 32 percent with no treatment to 5 percent with antibiotic treatment. And for patients aged 60 years or above, the risk of dying from pneumonia fell from an astonishing 62 percent with no treatment to 17 percent with antibiotic therapy. Similar spectacular reductions in mortality were seen in numerous studies focusing on antibiotic treatment of children with pneumonia.[64]

So, the answer was unequivocal: antibiotics do work for pneumonia. And therefore it is ethically unacceptable to give a patient with pneumonia a placebo. Period. It was thanks to due diligence and hard work by the FDA and IDSA personnel that this issue got resolved. But we should not forget that it was thanks to external political pressure on the FDA that the issue got raised in the first place.

We all want the FDA to continue to focus on its core mission, making sure that drugs approved for use in humans are safe and effective. And we are all disturbed when the rare scandal occurs in which drugs have slipped through the approval process that may not be as safe as we thought. But we must make sure that political pressure applied to the FDA is applied in areas that merit additional scrutiny, and that such pressure does not inadvertently squash any hope of developing lifesaving antibiotics that are desperately needed.

Several months after the IDSA-FDA pneumonia workshop, I went to Temple University in Philadelphia to visit my good friend, Dr. Bennett Lorber, professor of medicine and past chief of infectious diseases at Temple University Medical Center (not to mention an excellent guitar-playing musician and professional painter—a true Renaissance man). My trip to visit Bennett and Temple University Medical Center had nothing to do with the IDSA-FDA workshop, but during the course of our conversations, the topic came up. As it turned out, Bennett had long since familiarized himself with the old literature on antibiotic treatment of pneumonia. He had already read many of the original papers proving that antibiotics saved lives for pneumonia. We should have just asked him in the first place, and saved ourselves lots of trouble!

And Bennett told me something else that was extraordinary, related to the miracle of antibiotic effectiveness. Years earlier, the late Dr. Paul Beeson, who had been one of the most prominent American physicians of the twentieth century, had come to visit Temple University. During his visit, Dr. Lorber and Dr. Beeson had found themselves discussing the preantibiotic era and contrasting it with the antibiotic era. According to Dr. Lorber, Dr. Beeson related horror stories about what desperate physicians would do in the preantibiotic era when confronted with a life-threatening infection. For example, when confronted with a bacterial infection of the heart, Dr. Beeson specifically recalled using radiation therapy in a desperate attempt to kill the bacteria. This radiation "treatment" had no effect on the bacteria, did not cure the infection, and may have caused significant injury to the heart to boot. But, in the absence of antibiotics, what else could be done besides such desperate, primitive, and toxic measures when dealing with what was, at the time, a 100 percent fatal infection?

Dr. Lorber had the same reaction I had when reading the original studies of antibiotic effectiveness. The tone of writing in the manuscripts from the 1930s and 1940s indicated to us that the physicians clearly understood that they were on the cusp of nothing less than a revolution in medicine. It really was as if Edison was turning on the lightbulb. Suddenly, after three millennia, physicians had a real "magic bullet" that could be used to consistently, repeatedly, and reliably cure otherwise lethal diseases.

If you want to get a sense of how truly desperate and primitive medicine was in the era before antibiotics, consider the following. In 1937, Drs. Snodgrass and Anderson, working at a leading hospital in Glasgow, Scotland, published two of the first-ever clinical trials comparing antibiotics to a standard control therapy, for skin infections (so-called erysipelas).[65] I'd like to emphasize here that these clinical trials were at the cutting edge of medicine. I relate the history of these trials not to show you what medicine was like for the poor and underprivileged, receiving their care from charlatans in back alleys, but rather to illustrate what medicine was like for the well-off, receiving their care in highly respected medical institutions.

For their clinical trials, Drs. Snodgrass and Anderson compared the efficacy of the sulfa drugs, prontosil or sulfanilamide, to a standard control therapy, which was ultraviolet (UV) irradiation of the area of skin infection. Irradiation was used as the control therapy for the study because "it was already the routine method employed in the hospital." That is, UV irradiation was the standard therapy for skin infections because there was no other treatment to offer, and the physicians had to try something.

To maximize the rigor of their clinical trials comparing sulfa drugs to UV irradiation, Drs. Snodgrass and Anderson standardized all other aspects of their patients' care. The control and sulfa-treated patients were admitted to the same wards, were cared for by the same nurses and physicians, and were otherwise treated exactly the same, aside from the irradiation versus sulfa treatments. For example, as part of the standard background therapy, all patients were put on the same dietary regimen. The investigators wrote, "During the stage of [fever] the patients receive a fluid diet which may contain any of the following: barley water, Imperial drink, milk, glucose in water, Horlick's malted milk, chicken tea, arrowroot, cornflour, and junket." Of course, this diet had absolutely no medical benefits whatsoever, and frankly, it may have led to malnourishment by depriving people of solid food. Why would you have to be put on a liquid diet just because you have a skin infection and a fever? And why would that liquid diet have to consist of "Imperial drink" or "junket" (and what exactly *are* "Imperial drink" and "junket" anyway)? But, again, in the preantibiotic era, since they had nothing else that worked against disease, physicians felt like they had to do *something*. Fancy-sounding treatments like "arrowroot" and "junket" could provide psychological reassurance to the patients, and perhaps to the doctors themselves, that at least something was being done to intervene in the disease process.

Unfortunately, a weird and pointless diet wasn't the only standard "treatment" physicians used in that era. The authors continued: "On admission each patient was given a soap-and-water enema, repeated when necessary. Thereafter the same laxative—liquid paraffin—was used when required." So . . . the treatment for skin infections back then was a liquid diet followed by a mandatory enema, which was then typically repeated "when required." You may be wondering what effect an enema would have to treat a patient with a skin infection? Ah, that would be—none. Not one damn thing. But hey, as I said, the docs had to do something, otherwise why would people come to the hospital? And I guarantee you this, having a squirt gun jammed up your rectum and getting blasted up your colon with a high-velocity jet of soap and water ("liquid paraffin" anyone?) would certainly leave an impression with the patients that "something" indeed had been done!

Nowadays virtually everyone complains about their experiences in the hospital. They complain about the food, the loss of privacy, the impersonal nature of the care received, the frequent blood draws, the difficulty sleeping, and so on. But, when you get down to it, at least when you go to a hospital in the twenty-first century, you know that there are actual medical treatments that can be offered for most conditions. We can actually administer real therapies that really have an effect, rather than fake diets and pointless colon-blasting enemas. Patients prior to the 1930s had even worse food to eat in the hospital, and had no privacy whatsoever, being crammed into wards with a dozen or more people in one room and having their insides washed out every morning. To boot, they put up with all that, and in return got treated with methods that were completely and totally ineffective for their diseases.

We must not let ourselves forget the magnitude of the impact that antibiotics have had on the success of modern medicine. We should remember the contributions of those original pioneers who discovered and developed antibiotics. We must not take their gift to civilization for granted.

Mr. G. was a very nice man of about forty years old. He was a truck driver, who earned an honest living, was happily married, and had several children. He was busy going about his life, minding his own business, driving his truck on his route, when out of the blue, his back started to hurt. He thought it was just low back pain, which I gather is not an uncommon problem among truck drivers. The next day Mr. G. had a low-grade fever and continued back pain. He took an over-the-counter pain medication. He went to work. But, during the day, he began to feel weak in his legs. Now he became concerned. What was going on? His leg weakness was getting worse rapidly. Mr. G. decided to come to the hospital.

By the time he got to the emergency room, Mr. G. was virtually paralyzed below the waist. In the emergency room, a CT scan was done on his back. The CT scan revealed the presence of an abscess adjacent to Mr. G.'s spinal cord. Blood cultures were drawn and Mr. G. was started on antibiotics. He was rushed to the operating room, where the surgeons drained as much pus as they could to try to relieve pressure on Mr. G.'s spine.

I first met Mr. G. in the surgical intensive care unit after his surgery. He was awake but groggy from sedation. He was sweating profusely, both from his fever and tremendous pain in his back. I introduced myself as an infection specialist, and explained to Mr. G. that he had somehow gotten an infection that was attacking his spinal cord.

"Why did this happen to me?" he asked.

"I'm not sure. Have you fallen down or hurt your back recently?"

"No. I've been fine."

"Have you had any skin infections recently, or has anyone else in your family?

"No."

"Have you noticed any cuts or scrapes on your skin recently?"

"No, doctor, I've really been completely fine until two days ago. This just came out of the blue. I haven't been sick, and I haven't had any injuries."

"Has anyone you work with been sick or had any skin infections?"

"Not that I know of."

I shook my head. "Unfortunately, I can't explain why this happened to you. But we do see people come in with these types of infections and never figure out why they happened. The most important thing is to figure out what the bacterium is so we can give you the best antibiotics."

Mr. G. grimaced as he tried to move his legs but failed. "Am I going to be able to walk when this is all over?" You can tell when a patient is really terrified, like Mr. G. was. It was in his expression, his tone of voice, and his overall demeanor.

I paused for a moment, trying to figure out what to say. There was no right way to do this. It was important to balance not crushing his spirit with not giving him false hope.

"I'm not sure," I finally answered. "It's possible. We just have to hope that our antibiotics do the trick, and that the surgeons were able to get all the pus out. We'll get you working with a physical therapist. Just keep working at it, and we'll hope for the best."

Mr. G.'s face fell flat and he exhaled.

I told the same thing to Mr. G.'s wife, who was in the family waiting area. She started to cry and she glanced back at her two children, who were sitting with their grandmother. They were ten or twelve years old.

I knew what she was thinking. Who was going to pay the bills for the kids?

"Let's have a social worker come talk to you," I said. That was the best I could do.

Unlike Mrs. C., who had also had abscesses from MRSA near her spine, Mr. G.'s blood cultures never grew a bacterium. But the cultures from his spinal abscess did grow MRSA. How did the MRSA get to the spine? Where had it come from? We checked Mr. G. up and down, side to side, and back to front. But we never found where the MRSA had come from. Most likely, he had been carrying the MRSA on his body, with no symptoms. Then, he probably had some minor break in his skin that he never even noticed, and the MRSA managed to spill into his bloodstream. The bacteria then used the bloodstream as a superhighway, and got off at the exit near the spine.

Unfortunately our antibiotics just don't kill MRSA as well as they kill other *S. aureus*. Regardless, using a combination of several antibiotics and surgery, we managed at least to halt the progress of the infection. This took six weeks of intravenous antibiotics and a major surgery.

Is this a good outcome? Well, Mr. G. is alive, so that's good. And he regained a little bit of strength in his legs. But only a little. I met him again a few months later. He was at a rehabilitation hospital. When I walked in the room, Mr. G. was lying in bed.

"How are you?"

"I'm okay," he answered, but he was clearly depressed.

"How are the legs doing? How much can you move them?"

He grabbed both sides of his bed to brace his upper body as he slowly moved his legs side to side across the bed. This was improved from previously, a little.

"Can you lift them off the bed?"

He shook his head.

"Give it a shot. Let's see how well you do."

He again braced his hands against the sides of the bed. This time he grunted and grimaced as he struggled to lift his legs up. They made it perhaps an inch or two above the bed, and he was able to hold them up for a few seconds before they fell back down. He was breathing hard from the effort.

"That's better!" I tried to encourage him.

He offered a half-hearted smile. "Yes, I'm still working to get better."

I tried to follow Mr. G.'s progression even though I was no longer involved in his care. Last I heard, which was perhaps six months after his initial infection, Mr. G. still could not walk. Unfortunately, he had developed a massive bedsore. He obviously could not work. He did not know how he was going to support his family.

Mr. G. hadn't known about the problems of antibiotic resistance when he came to the hospital. When he left the hospital, all he knew was it would have been nice if the antibiotics we gave him would have worked better. If so, maybe he would be walking now and working. Mr. A., Mrs. B., Mrs. C., Mr. D., Mr. E., and Mr. F. also could have used new antibiotics. So could Rebecca Lohsen, Carlos Don, and Ricky Lannetti, and the other unfortunate patients whose stories are told on the IDSA's Web site. So could our troops in Iraq and Afghanistan, who, as if to rub salt into their wounds, are trying to overcome drug-resistant infections in combat wounds without the antibiotics they need. All of these, and the hundreds of thousands of Americans per year that contract antibiotic-resistant infections, are counting on society to come up with a way to reinvigorate antibiotic development. It's up to us.

So, how do we do it?

CHAPTER 7

forget pharmaceutical companies, the government can create new antibiotics—not!

Clearly, in the long run, we need to have new antibiotics developed so we can keep treating lethal infectious diseases. Having accepted this premise, the next question becomes, how should we discover and develop these new antibiotics?

Due to the prices of prescription drugs, as well as aggressive government lobbying by the pharmaceutical industry, there has been a massive backlash against pharmaceutical companies in the media and the public. As a result, the notion of creating legislative incentives to stimulate new drug development by pharmaceutical companies is less than popular, to say the least. The idea I most commonly hear expressed about how to get around the need for legislative incentives aimed at pharmaceutical companies is to have the government be directly responsible for discovering and developing new antibiotics.[1] Certainly it is true that the government is responsible for resolving critical societal problems that affect the general public. But the devil is in the details. How should the government go about directly developing new antibiotics? There are three versions of this idea, each of which is problematic.

CREATE A NATIONALIZED DRUG DEVELOPMENT CORPORATION

One idea is to create a national pharmaceutical company funded and controlled by the government that directly carries out drug discovery and development. In this way, society can exert precise control over which drugs are developed. While this idea may sound practical, one has to seriously question how realistic it is to believe that a not-for-profit, government-managed entity is going to be able to discover, develop, and commercialize new antibiotics.[2]

The US FDA Web site has an electronic database called the "Orange Book"—http://www.fda.gov/cder/ob/. The FDA Orange Book is a record of all drugs approved for use in humans in the United States, and contains data on which company or manufacturer submitted the New Drug Applications that were ultimately approved by the FDA. A New Drug Application is the formal documentation submitted by a drug manufacturer to request FDA approval so a drug can be commercially marketed in the United States. A New Drug Application contains all of the available information about the approved drug, from its chemistry and manufacturing, to efficacy and safety data from animal studies, to all of the efficacy and safety data from clinical trials in patients. The FDA reviews the New Drug Application and decides, based upon material contained within it, if sufficient safety and efficacy data are available to warrant approval to begin selling the new drug to consumers. Hence, all drugs marketed in the United States must have had a New Drug Application filed with and approved by the FDA, and the Orange Book is the record of all of these approvals.

To evaluate the contributions, to date, of government-sponsored drug development to the current antibiotic armamentarium, I created a list of twenty-first-century antibiotics by recording every antibiotic discussed in a standard infectious-diseases textbook.[3] I then evaluated who brought these drugs to market by reviewing the data on each drug in the Orange Book.

It turns out that of 114 antibacterial antibiotics that were ultimately approved by the FDA for use in the United States, 109 (96 percent) were exclusively developed and brought to market by giant pharmaceutical companies. Of those 114 antibiotics, 21 have been withdrawn from the market or are no longer manufactured, and 93 have active New Drug Applications (i.e., are still available in the United States). Of those 93 currently available antibiotics, 90 (97 percent) were exclusively developed and brought to market by giant pharmaceutical companies.

The five antibiotics not developed exclusively by giant pharmaceutical companies are:

(1) Daptomycin, which was ultimately brought to final approval by the biotechnology company Cubist. However, the giant pharmaceutical company Lilly discovered daptomycin and spent more than fifteen years developing it before they out-licensed it to Cubist, which completed the final step in development and filed the drug's New Drug Application;
(2) Gemifloxacin, which was brought to market by the biotechnology company Oscient, but only after being discovered by the giant multinational company LG Chemical Investments Inc. The drug's multimillion-dollar Phase II and Phase III clinical trials were then paid for and completed by the giant pharmaceutical company SmithKline Beecham (prior to its merging to become GlaxoSmithKline), which later divested its ownership in the drug;
(3) Sparfloxacin, which was developed and initially taken through FDA approval by the giant European pharmaceutical company Rhone-Poulenc Rorer (prior to its merger with Hoescht to form Aventis), and was then licensed to the biotechnology company Mylan in 1998, and was subsequently taken off the market;
(4) Cefditoren and (5) Cefmenoxime, both of which were developed by the company TAP, which is a joint venture of the giant multinational pharmaceutical companies Abbott and Takeda. Cefmenoxime is no longer available in the United States.

Hence, 114 of 114 antibacterials reviewed were developed with extensive major pharmaceutical company participation. In summary, of the approximately one hundred antibiotics that are currently available for use in the United States, not one has been developed by a government-run entity. Not one. Every single antibiotic available in the United States was discovered, developed, and brought to market by for-profit corporations. This track record does not bode well for the future of drug development by a theoretical government-run pharmaceutical company.

Furthermore, I am hard pressed to identify a single antibacterial agent in use anywhere on Earth that has been developed by a government-run entity. The only anti-infective in the world of any type that was exclusively developed by direct government sponsorship is the anti-malaria drug artemisinin (which is not available in the United States). Artemisinin, or *quinghaosu* in Chinese (meaning "extract of green herb"), was developed as a modern drug by researchers working for the Chinese army.[4] Artemisinin derives from a plant called sweet wormwood, or *qinghao* in Chinese (meaning "green herb"). Chinese herbalists have been using *qinghao* to treat malaria for millennia.[5] So, it is hardly fair to say that the Chinese army discovered the drug. Rather, researchers working for the army developed techniques to mass extract and produce the drug from the source plant, *Artemisia annua L.*

Other antimalarial agents, such as quinine and chloroquine, were developed as drugs either exclusively by large pharmaceutical companies, or by partnerships of US and European militaries in conjunction with large pharmaceutical companies.[6] But again, these drugs were developed from plants or herbs that had already been widely used since at least the seventeenth century to treat malaria,[7] so there was no real drug discovery involved. The work that had to be done was largely extraction of the active compounds and engineering new processes to manufacture or harvest the compounds. Yet still US and European militaries relied heavily or exclusively on giant pharmaceutical companies to carry out this work.

Unfortunately, the issue of government sponsorship of drug discovery is emotionally charged, because advocates of the idea so despise the concept of financial incentivization for private pharmaceutical companies. But, whether we like it or not, for better or for worse, the reality is that we live in a capitalist society. I neither applaud nor criticize this fact, I merely state it as a fact. Historically, government-run corporations have been inconsistent with the manner in which our society operates. This of course does not preclude the establishment of the first such entity in the United States designed to discover and develop drugs. But one has to seriously question the feasibility of the creation of a competitive corporation owned and operated by the federal government, especially given the track record of success in drug development of private pharmaceutical companies, and the complete absence of such a track record for a government pharmaceutical company.

There are government-run drug companies in other countries (mostly underdeveloped countries), but their primary role is to create cheap, generic copies of drugs that were discovered and developed by private companies. They do not discover and develop new drugs. While such government-run drug companies may be good at copying other companies' drugs at a cheaper price, they are notoriously incapable of innovation, or of novel drug discovery and development. To reiterate, there are no examples of any antibacterials in use anywhere in the world that have been developed by such entities. Nor am I aware of examples of any other types of drugs, of any class, that were developed by government entities. Private pharmaceutical or biotechnology companies have exclusively developed virtually all drugs of all types (not just antibiotics) that have been approved for use in the United States, with only minor contributions of molecules developed out of academia or government in partnership with private companies.[8]

WHAT DO YOU MEAN GOVERNMENT-RUN ENTITIES CAN'T DEVELOP DRUGS? SHOULDN'T OUR GOVERNMENT-SPONSORED SCIENTISTS DO THIS WORK?

The US National Institutes of Health (NIH) is by far the largest government-funded scientific operation in the world. As of fiscal year 2008, the overall budget for the NIH was $28.9 billion.[9] The budget request for the NIH for fiscal year 2009 is nearly identical ($29.2 billion). Given this budget, it is not surprising that people commonly express outrage that pharmaceutical companies reap the benefit (i.e., profit) from drugs that are developed built upon a backbone of basic scientific research sponsored by such a substantial investment by US taxpayers. The question must be asked, why can't the NIH use its $29 billion budget to directly develop new antibiotics, rather than relying on private companies to do the job?

The NIH does not—I repeat, does not—do drug development. Period. The NIH funds basic science research. Yes, it is true that basic science discoveries provide the foundation for future potential drug discovery. But no, it is not true that these discoveries can be developed into drugs and transferred into the commercial realm by academic, basic scientists using NIH funding.

We have previously discussed how important basic science research is to lay the foundations for future technology development. Excellent examples of drugs developed by building on a foundation of basic science knowledge are anti-HIV drugs, as discussed in chapter 6. But the key concept is, once the basic science knowledge was in place, all of the actual technology (i.e., the anti-HIV drugs) was created and developed by private corporations, *not* basic scientists funded by the NIH. While many such drugs have been developed by companies that relied upon a foundation of basic scientific knowledge laid by government-funded scientists, virtually no drugs used in humans have ever been developed by solely using NIH funding.[10]

The NIH funds science, not technology. Science is a method built upon hypothesis testing, which is used to ask basic questions to explore how the universe works. Avoid confusing science and technology, because in the real world, there is a huge difference between them, especially in biology and medicine. Biomedical scientists in basic research are not funded by the NIH to develop applied technology. They are funded to make scientific discoveries that explain how things work. The term *mechanism* is a key buzz word in NIH grants. The goal of NIH-funded research is to identify the "mechanisms" by which the natural world works. Technology experts, who are a completely different set of people from basic biomedical scientists, can then apply knowledge about how the universe works to develop useful technology, or products.

To distill the difference between science and technology down into the world of medicine, NIH-funded academic scientists make basic discoveries about how disease occurs. But academic research laboratories lack the expertise, the access to the enormous capital (up to one billion dollars per drug that is ultimately approved),[11] the manpower, and the equipment necessary to actually develop drugs to treat those diseases. My colleagues and I have considerable personal experience with this. My NIH-funded academic laboratory and those of my colleagues are focused on understanding how the fungus *Candida* and the bacterium *S. aureus* cause disease in humans, and how the human immune system works to fend off those microbes. Collectively, *S. aureus* and *Candida* cause more than 150,000 bloodstream infections per year in the United States, accounting for more than 40,000 deaths.[12] You can see why it might be nice to have a vaccine that could prevent these infections. By doing the basic science work, and learning how the interaction between host and microbe leads to cure or disease, we have hoped to lay the groundwork for just such a vaccine.

For example, in developing our vaccine against *Candida* and *S. aureus*, we first had to understand how the microbes cause disease in humans (step 1 = basic science discovery). We also had to understand how people normally defend themselves from infections caused by these microbes. In the course of these basic science studies, we identified a family of proteins from *Candida* that was critical to the fungus's ability to infect human tissues, and we found that the fungal protein was shaped similarly to a family of proteins from *S. aureus*.

We then began step 2, translating the basic science discovery into a candidate molecule for development into a vaccine. Our laboratory is one of a relatively small but growing number of academic biomedical laboratories in the United States that is focused on bridging the gap between strictly basic science and strict technology development. We are referred to as "translational" scientists, because our goal is to translate basic science discoveries into practical treatments for disease. Fortunately, under its current leadership, the NIH and its NIAID division—the division focused on infectious diseases—have recognized the critical need for additional translational research in the United States. The NIH has begun to place a greater emphasis on funding translational research projects, and the NIAID has begun to put together programs and build infrastructure to help translate molecules in academic laboratories and into real treatments. Nevertheless, it still can be difficult to secure funding for such projects.

So, to pursue translation of our potential vaccine into a candidate molecule, we produced copies of the *Candida* protein in the laboratory and proved that vaccination with this protein protected mice from infections caused by both *Candida* and *S. aureus*. These translational science studies involved testing different doses of vaccine, different routes of administration, mixtures in different types of enhancing solutions, and so on, in order to identify the vaccine formulation that was most effective in mice. We also tested the vaccine against many different strains of fungus and bacteria to see how broad the protection was, and we tested how fast the vaccine worked and how long its effect lasted. Virtually none of these experiments was of much interest in the basic science world, because they did not explore the mechanism by which the vaccine worked. Nevertheless, from a translational-science/technology-development perspective, they were absolutely critical. Ultimately, we developed a vaccine regimen that protected mice from lethal infections caused by *S. aureus* and *Candida*.[13]

And that is where our ability to develop the vaccine further hit a brick wall. As I said earlier, no academic laboratory has the access to capital or expertise required to complete the third step of biomedical technology development, technology transfer (which I will loosely define as completion of the steps required to transfer a candidate molecule out of the laboratory and turn it into a commercialized technology). Technology transfer must be accomplished with corporate involvement. In fact, the mission of the NIH is to fund basic science that establishes a foundation upon which biomedical technology can be transferred to the commercial realm. Critics who complain that the taxpayer-funded NIH subsidizes drug discovery by pharmaceutical companies are actually being ridiculous. The NIH is *supposed* to subsidize industry development of technology that can be made commercial by funding basic science discoveries that open avenues for new technology. It is, in fact, the very mandate of the NIH to lay the basic science groundwork atop which others actually develop technology. Indeed, if basic science discoveries funded by NIH were not turned into useful technology by companies, then why would we be doing the research in the first place? Science for science's sake is a noble concept, but it is not practical in a setting where public taxpayer dollars are actually paying for the science. Taxpayers have a legitimate expectation that lifesaving medications will be developed as a result of the billions of dollars of their money that goes to support basic biomedical research.

So now, after translating our basic science into a vaccine candidate, our group has had to proceed to step 3, technology transfer, which is the development of our vaccine into a product that can be sold commercially and used clinically. In order for us to achieve technology transfer, we had to form a startup biotechnology company. That company has had to raise the enormous amounts of capital (in the multiple millions of dollars) required to affect even early phase drug/vaccine development. The company must also supply myriad expertise that the academic scientists are incapable of providing. Such expertise lacking in academic laboratories includes: (1) legal council to develop intellectual-property protection (without which the entire exercise is moot); (2) business expertise to negotiate contracts and terms, establish a corporate structure, acquire research space, and establish requisite ancillary services (such as picking up lab wastes, meeting California and federal occupational work and safety requirements, creating a supply purchasing and delivery mechanism, establishing human resources including employee insurance and benefits, establishing accounting services, etc.); (3) technology development expertise; and (4) regulatory expertise.

To give you a clearer picture of how high the barriers are to successfully create new technology, let me expand briefly on the expertise required to achieve technology development and to meet government regulatory requirements. The technology development expertise that is necessary to make a new drug or technology includes things like establishing methods to manufacture the product to meet FDA standards. The FDA requires that all products eventually tested in humans must be manufactured by so-called Good Manufacturing Practice, or GMP. GMP-compliant manufacturing ensures that the drug or vaccine is pure, free of toxins, and is reproducibly manufactured each time a batch is run.

Achieving GMP is an incredibly complicated process that involves molecular biology, chemistry, bioengineering, and quality assurance/quality control. In addition to certifying the "seed lot" from which all subsequent batches of product will be made, the GMP process requires actually inventing a new manufacturing process, step by step. All materials used in the manufacturing process must be certified and determined to be pure and free of potential toxins. Levels of contaminants must be actively monitored, and purity of product must be confirmed. Batch-release testing must be performed to ensure that each lot of the product is the same as the prior one. Stability testing must also be performed to define for how long the product is stable, whether it must be stored in the refrigerator, and/or whether it can be frozen for long-term storage. The GMP process must be done by highly experienced personnel and in a building that meets FDA standards for bioengineering (e.g., there are standards for maximum particles per cubic centimeter of air allowed in processing rooms, there have to be dedicated water lines with specific purity parameters, etc.).

In general, it can take twelve to eighteen months to establish GMP for a product, or potentially even longer. As far as cost, we've seen GMP bids of up to $3 million for our vaccine, and have heard that it can be much more expensive if the production process is more complicated. I can assure you that the above expertise and infrastructure is absent in every academic lab in the country.

It turns out that the NIH is aware that accomplishing GMP is a major barrier to translation of promising treatments from the academic laboratory to the patient's bedside. Indeed the NIH has developed its own GMP-compliant manufacturing facility at its Vaccine Research Center.[14] That facility is currently enabling vaccines that are internally developed at NIH to undergo GMP-compliant manufacturing without initially requiring a corporate partner. However, the facility is not available to scientists who work outside the NIH, even if those scientists are currently funded by grants from the NIH. Furthermore, even for vaccines developed internally at NIH, the very mission statement of the vaccine center makes clear the need to eventually involve private corporations after the manufacturing stage, in order to develop viable, commercial technology. Here is a direct quote from the NIH Vaccine Research Center's mission statement: "The VRC will actively seek industrial partners for the development, efficacy testing, and marketing of vaccines."[15]

Other options for completing the manufacturing of technologies that derive from NIH-funded research do exist. However, the magnitude of manufacturing resources is limited compared to the sheer volume of candidate molecules awaiting development. Availability at NIH is therefore very scarce. For instance, it was only through more than a year of persistent effort by two dedicated NIH officials that the vaccine our group has been developing has finally been able to access such resources, and enabled us to manufacture the vaccine candidate to FDA specifications.

The NIH is continuing to attempt to expand its infrastructure to help bridge the gap between science and technology. Indeed, much progress has been made in the world of therapeutics, to help scientists in the lab translate their science into useful technologies. And the NIH's efforts in this regard should be encouraged and strengthened. But it is a plain fact that all of the NIH's efforts are designed to help scientists connect with private capital, which is essential to enable completion of technology transfer. There is absolutely no model whatsoever for development of technology solely by NIH, without participation of private corporations.

As I mentioned before, to transfer new treatments from the laboratory bench to the patient's bedside, one also needs to have a complete familiarity with the federal regulations governing drug development and marketing. Specifically, one needs to understand government requirements for establishing safety and efficacy of the product in animals before human testing is allowed. Ultimately the Investigational New Drug (IND) application that is submitted to the FDA to seek permission to test a product in humans can be hundreds of pages long. Such IND applications include data on safety and efficacy of the product in animals, GMP, shelf-life of the product (i.e., product stability), and everything in between. If your company has spent equity or put up collateral to raise the enormous capital required to generate data for an IND, you sure don't want the FDA to reject that IND application. If the FDA does reject the IND, a company that has spent millions of dollars and many years pulling together all the information to include in the report is likely to find itself in major trouble with its investors. That company is not likely to be around for very long. So you'd better be working with an expert who has lots of experience in putting together IND applications. Again, this expertise is lacking in academic laboratories, and this expertise costs significant capital to hire.

As you can tell from this partial description of the steps necessary to effect technology transfer of a drug or vaccine, the cumulative expertise and capital required are far beyond the capacity of any academic laboratory. As I alluded to earlier, the NIH does participate in government programs that fund small businesses to partner with academicians to transfer technology. These small-business grants can be very useful to kick-start the process. The standard small-business grant supplies a maximum of $1.65 million over a three-year period. Recognizing that this amount of money pales in comparison to what is actually needed to affect technology transfer, the NIAID (again, the division of the NIH focusing on infections) has created newer "Advanced Technology" grants to support vaccines and other cutting-edge immune-based treatments for infections.[16] These Advanced Technology grants supply a maximum of $3.6 million over five years in two separate phases (a first phase of up to $300,000 per year for up to two years, and a second phase of up to $1 million per year for three years). But, do not be fooled. In the grand scheme of things, $3.6 million is still a drop in the bucket compared to what is required, and spreading the funding out over three to five years is totally inconsistent with the needs of technology development. Ten to a hundred times as much, or more (possibly *much* more), is needed to complete the process of bringing a product to market, and this money is typically spent in a very focused period of time (e.g., it is needed over a span of eighteen months in our case).

The NIH makes explicitly clear in its instructions for submitting small-business grants that the NIH expects to provide only seed funding, and expects academicians to partner with companies to subsequently raise the much larger amounts of private capital that are required to develop new drugs and other treatments. The NIH refers to this period of private capitalization as "Phase III" of the development of the treatment. Here's a direct quote from the grant instructions: "An objective of the SBIR/STTR program is to increase private sector commercialization of innovations derived from Federal R/R&D. During Phase III, the small business concern (SBC) is to pursue commercialization with non-[grant] funds. . . . Phase III is the period during which Phase II innovation moves from the laboratory into the marketplace. No [grant] funds support this phase. The small business must find funding in the private sector or other non-[grant] Federal agency funding."[17]

All of this brings me to a final point. As I have emphasized, the belief that NIH funding of basic science work constitutes the majority of funding required to discover and develop new drugs is absolutely, flat-out wrong. In fact, collectively, far more private capital is spent by biotechnology/pharmaceutical corporations every year on research and development than the NIH spends.[18] In 2004, the research and development budgets of just the ten largest pharmaceutical companies exceeded the NIH's annual budget.[19] Add in the dozens of other pharmaceutical companies, and hundreds of smaller biotechnology companies all over the country and throughout the world, and spending of corporate capital and venture capital money massively exceeds taxpayer-funded bioscience research. It has recently been calculated that the collective US corporate biopharmaceutical research and development expenditures for 2006 amounted to $55 billion, almost twice the NIH budget.[20] In 2007, that number increased to $59 billion,[21] which was more than twice the NIH budget of $28.6 billion.[22] Furthermore, in contrast to the NIH, this private capital is spent with a focused goal of specifically bringing products to market.

So, the idea that NIH-funded—or other government agency–funded—scientists are going to develop and commercialize new antibiotics is borne of a lack of understanding of the very real, very palpable difference between science and technology, as well as a fundamental misunderstanding of the mission of the NIH. Private corporations are going to have to be involved in the process.

GOVERNMENT CONTRACT

Okay, fine, so a government-run entity is unlikely to be able to discover and develop new antibiotics, and taxpayer-funded basic science research is not going to enable translation of science into commercial technology without corporate involvement. No problem. Just have our government put out contracts to private corporations, in the same way that the government puts out contracts to the defense industry for a new battle tank or a new fighter plane. That is, have the government create contracts to pay companies to develop needed drugs and possibly even provide a guaranteed market for purchase of such drugs.

I used to think this was the way to go. Then I met Chuck Ludlum, who at the time was a senior staffer in the office of Senator Joseph Lieberman (Chuck subsequently retired from the political arena and went to work for the Peace Corps in Africa—go Chuck!). Chuck—who is about the smartest person I've ever met when it comes to understanding the business and politics of healthcare—explained to me and my colleagues the ways of the world.

Pharmaceutical companies use an entrepreneurial business model. They expect to take all of the risk in drug development and pay all of the research and development costs. If they invest $500 million in a drug that eventually fails in phase III clinical trials, it's their loss. Pharmaceutical companies also face the challenge of constantly being on a treadmill because of the threat of generic competition. A pharmaceutical company can invest hundreds of millions of dollars in developing a new drug, and then the company has about twelve years, on average,[23] to recoup its investment before the remainder of the patent on that drug expires, and generic companies come along and copy that drug. There is no guarantee that pharmaceutical companies will be able to replace currently on-patent drugs with new drugs as the current drugs go off patent. In contrast, other types of manufacturers can continue to sell products at high volume, with low margins, regardless of patent status. For example, toilet paper manufacturers probably don't spend a lot of time worrying about completely replacing their toilet paper technology every twelve years.

In return, to offset this increased set of risks, pharmaceutical companies require a higher profit margin on the sale of their products than many other industries. Indeed, in 2004, *Fortune* magazine found that the pharmaceutical industry ranked third among all industries, with a 14.3 percent profit margin (i.e., the ratio of profit to overall sales).[24] Without this high profit margin, the increased risk assumed by pharmaceutical companies would make continued investment in their business impossible.

This entrepreneurial model used by pharmaceutical companies is precisely the opposite of the business model used by defense contractors. Defense contractors have most of their research and development costs paid for by government contracts. Furthermore, their products have such complexity of systems integration, as well as classified technology, that there is no issue with a generic company coming along and copying the technology or the end product after patents expire. For example, no generic manufacturer is going to be capable of coming along and making generic, cheaper copies of an F-22 Raptor fighter plane or an M1 Abrams tank. This business model allows defense contractors to assume smaller up-front risk during the development period. This business model also allows defense contractors to continue to profit from successful products for decades, without facing cheaper, generic competition. Because defense contractors assume a much lower risk during the research and development phase, and can profit from their products for longer than pharmaceutical companies, they accept a much lower profit margin if and when the product is ultimately made available for commercial sale. Indeed, according to *Fortune*, in 2004, the aerospace and defense industry ranked thirtieth overall, with a 3.2 percent profit margin.[25]

So, pharmaceutical companies, in contrast to defense contractors, do not operate under a business model that is amenable to government contracts. In fact, their business model works the opposite way. Nevertheless, just because that's the way they have done business for a hundred years doesn't mean they can't change, right? Well, in fact, this idea has been tested.

The 108th US Congress passed legislation called Bioshield. Then President Bush, who was a staunch advocate of the bill, enthusiastically signed it into law in 2004. Bioshield has made available almost $6 billion, mostly in the form of federal contracts, for pharmaceutical and biotechnology companies to develop drugs, vaccines, or other "countermeasures" for biological, chemical, nuclear, and radiological weapons. This bill does precisely what we've just been talking about. It replaces the standard entrepreneurial business model that dominates in biotechnology and pharmaceutical companies with a defense contractor model. The purpose of Bioshield is to create an incentive for these companies to focus on making new treatments for bioterrorism pathogens, despite the fact that there is no commercial market for such products.

So, how has Bioshield worked out?

"Bioshield has failed miserably." That's a direct quote from a former senior official with the Department of Health and Human Services.[26] Darren Fonda of *Time* magazine wrote: "BioShield hasn't transformed much of anything besides expanding the federal bureaucracy. Most of the big pharmaceutical and biotech firms want nothing to do with developing biodefense drugs."[27] Similarly, according to Marc Kaufman of the *Washington Post*, "Despite . . . passage of the . . . project BioShield bill designed to speed development of new products, officials say the nation is scarcely any better protected than it was in 2000."[28]

Of note, one product that has been developed as a result of Bioshield was a second-generation vaccine against smallpox, which is intended to replace the previous, first-generation smallpox vaccine. While it is theoretically possible that terrorists could launch a smallpox attack in the future, there is no evidence that terrorists have access to smallpox, or that they have the expertise to weaponize it.[29] There may have been paranoia about the potential for a smallpox attack based on the same faulty evidence that the Bush administration used to justify invading Iraq (i.e., that Saddam Hussein had weapons of mass destruction, including bioterrorism weapons, which was subsequently found to be false).[30] But in the end, the scare over smallpox was theoretical and not based on any reliable information.

Furthermore, I wish to emphasize that *smallpox does not exist in nature anymore*. It has been eradicated from the face of the Earth. The virus that causes smallpox exists only in two research stockpiles, one kept by the US government and one kept by the Russian government.[31] The last case of smallpox on Earth occurred thirty years ago (in 1978, and it was acquired due to a research laboratory accident)—the last naturally acquired case on Earth occurred in Somalia in 1977, and the last case in the United States occurred more than a half century ago, in 1949.[32]

Furthermore, the reason why smallpox no longer exists in nature is that the first-generation smallpox vaccine eradicated it. In fact the first-generation smallpox vaccine is the only vaccine in history to totally eradicate its target disease from nature. It is, quite arguably, the most successful vaccine ever invented.

Under Bioshield, the US government spent $429 million[33] to support efforts by a biotechnology company to develop a new vaccine to replace a previous vaccine that has been arguably the most successful vaccine in history. The nearly half-billion dollars our government spent on a vaccine that was quite arguably not needed is tenfold more than the amount of money the government spends on research for antibiotic-resistant bacteria, which actually do kill tens of thousands of Americans every year. Furthermore, after supporting the development of the new vaccine, the government then provided a guaranteed market for the new vaccine. That is, they promised the biotechnology company that they would purchase tens of millions of dollars worth of the new vaccine, since they knew no one else would, because there is no smallpox in nature.

The US government has spent hundreds of millions of dollars to buy this new vaccine despite the fact that we still have 75 million doses left of the first-generation smallpox vaccine, which is likely more than would be required in the highly unlikely event of a terrorist attack. So, the real question is, was it really wise of the US government to spend nearly half a billion dollars to enable a biotechnology company to profit from the development of a new vaccine against a disease that no longer exists (and hasn't existed for decades) because it was eradicated by the prior vaccine? Wouldn't it have been better to spend those many millions of dollars to pay other companies to develop technologies that could prevent or treat real diseases, which actually exist in nature and actually kill tens of thousands of Americans per year?

Not only does Project Bioshield cut precisely against the grain of the standard business model under which pharmaceutical companies normally operate, but, believe it or not, many companies had horrific experiences working with the federal government during the Bush administration. I realize that many people think the Bush administration was in the pockets of the pharmaceutical companies, but let me assure you that during his administration's tenure, there were well-publicized actions taken by the federal government that literally scared off both large pharmaceutical and small biotechnology companies from going anywhere near a government contract.

For example, in the aftermath of the anthrax scare in 2001, the Bush administration threatened to revoke Bayer's patent on ciprofloxacin if the drug maker did not supply an enormous quantity of the antibiotic at half of the standard cost.[34] You'll recall that the 2001 attack involved someone mailing anthrax spores to offices of people in the news media and in Congress. The anthrax scare resulted in a national hysteria that was enormously out of proportion to the five people who were tragically killed during the incident. What was needed at the time was a composed and rational leadership that would have helped calm the waters, provide reassurance of the very small scale of the attack, and help decrease the public demand to hoard antibiotics as a panic response to the attack. Unfortunately, the Bush administration delivered the exact opposite. They fueled the fire further by releasing press statements indicating future attacks were likely, and these press statements resulted in a public panic that overwhelmed affected health officers and laboratories.[35]

The public panic inflamed by the Bush administration also drove the desire to increase the availability to the public of a very powerful, very precious antibiotic, ciprofloxacin. Ciprofloxacin and its close relative, levofloxacin, are the only oral antibiotics that can treat many drug-resistant gram-negative bacteria. We desperately need to preserve these antibiotics. Preservation does not include doling them out like candy to every citizen who is terrified by a heinous anthrax attack. But as a result of the Bush administration fanning the flames of paranoia, the anthrax attack caused ciprofloxacin prescriptions to increase by an astonishing 160,000 in 2001 versus 2000.[36] Virtually that entire increase was due to prescriptions written for panicked citizens who were in no danger whatsoever from anthrax.

The poor judgment and questionable constitutionality of the Bush administration's threat against Bayer's patent aside, the very act of making the threat had a chilling affect on corporate interest in accepting contracts from the federal government. Bayer had no direct link to the federal government, and ciprofloxacin had not been developed using federal grants or contracts. It had been developed by Bayer, using Bayer research and development dollars. Yet still the federal government was threatening the core of the company's intellectual-property protection unless Bayer knuckled under. Now imagine you are a different pharmaceutical company and you are considering whether you should develop a product under a federal contract. What effect do you think the Bayer incident would have on your decision making? How much worse would it have been for Bayer had they actually developed ciprofloxacin through a contract from the federal government? They likely would have been forced to give over the entire stockpile at no cost at all, and had no constitutional protections against such a seizure of inventory. In the aftermath of this debacle, pharmaceutical companies do not want to subject themselves to the inherent strings attached any time government money is involved.

But the Bayer story isn't even the most egregious interaction between the Bush administration and the pharmaceutical world. Take the example of Hollis-Eden Pharmaceuticals. (I should disclose here that I personally met with executives at Hollis-Eden about their research portfolio, and I received a $1,500 consulting fee from them in 2002.) Hollis-Eden has been developing a highly promising drug that could ameliorate the effects of radiation poisoning in case of a nuclear or "dirty-bomb" attack. Working under the auspices of Bioshield, Hollis-Eden invested $100 million in developing the drug based on an understanding that the federal government anticipated ordering ten million doses of the drug for military and other first-responder (i.e., paramedics, emergency physicians, etc.) use. After the drug development was done, the Bush appointee in charge of the Bioshield program (whose qualification to be a scientific administrator was that he had previously been a lawyer for Amtrak) arbitrarily decided to order only one hundred thousand doses of Hollis-Eden's drug—that is, one hundred times fewer doses than originally promised.[37] Hollis-Eden's stock went into a tailspin when that news was announced.

But it got still worse for Hollis-Eden. Even more recently, the government decided to cancel entirely its original request for the drug. How did Wall Street respond to this totally capricious decision? Yet another 32 percent drop in market capitalization for the victim, Hollis-Eden, which made the mistake of trusting the federal government at its word. As one Web-based news service reported, "Government Nukes Hollis-Eden's Radiation Drug."[38]

In the aftermath of these screwups, blunders, and mishaps, what do you think the probability is that any form of government contract is going to be able to lure pharmaceutical companies back into antibiotic development? Clearly other solutions to the problem are needed.

CHAPTER 8

"toxic pharmaceutical politics" and finger pointing

Before considering solutions that have a chance to restimulate antibiotic development, we need to consider one more barrier to antibiotic development. We have discussed why decreasing antibiotic use, while very important to slowing the spread of resistance, is ultimately not going to solve the problem of pandemic, antibiotic-resistant infections. We need new drugs to allow us to keep pace with infections caused by constantly evolving microbes. We have also seen that government-run entities, taxpayer-funded basic science, and government contracts are unlikely to be capable of reinvigorating the discovery and development of new antibiotics. So if we need new antibiotics, and the government is not going to discover and develop them, it becomes clear that, ultimately, some form of incentive is going to have to be created to rekindle the interest of private corporations in antibiotic development.

But there is a giant problem with creating such incentives. Pharmaceutical companies have an abysmal reputation among the public and politicians. In a recent book-turned-movie, *The Constant Gardener*, pharmaceutical companies were even compared unfavorably with arms dealers! Pharmaceutical companies are frequently blamed for the rising cost of healthcare. They are excoriated for charging inordinate prices to senior citizens and sick people, and preying upon the illnesses of these victimized populations to extort money from them. They are decried for aggressive advertising and marketing of their drug products to physicians and directly to consumers. Pharmaceuticals are the companies everyone loves to hate. So, it is not shocking that the very mention of financial incentives for pharmaceutical companies is enough to send many politicians diving for cover, and to raise the ire of the public and of physicians.

Unfortunately the pandemic of antibiotic resistance is not going to disappear just because society currently lacks the political will to address it. So, we are going to have to make a choice: (1) Give up on antibiotic development. The consequence of this choice could be a return to the preantibiotic era for many types of infections; (2) Invent a completely new mechanism to discover and develop new antibiotics. I'm certainly willing to listen if you have some ideas on how to do this; or (3) Objectively consider all the facts, remove all of the emotional baggage, and find palatable, financially responsible ways to provide incentives to stimulate pharmaceutical and biotechnology companies to discover and develop new antibiotics.

MY DISCLOSURES

Before we proceed, you need to know some things about me. I need to tell you how I make my living, because I do have potential conflicts of interest when it comes to the issue of creating incentives to stimulate antibiotic research and development. It is up to you to factor in my conflicts of interest and decide for yourself how to interpret my comments in light of these conflicts.

I am an academic physician-scientist. Obviously the physician part means that I have an MD, and I do see patients. However, I do not see private patients. I see only patients who are cared for at the public county hospital where I work. In this setting I serve in a supervisory capacity for students, residents, or subspecialty fellows rather than as a primary care physician. I do not bill patients, and I make no money from the patients that I see or from their insurance companies (for the few who have insurance).

Furthermore, as an academic physician, I do not get promoted through the faculty ranks based on my patient care activities. Rather, my promotions are primarily based on securing grant funding to further my research. Not only are my promotions based on how much research grant funding I get, but to date my paycheck has been as well. While this is in the process of changing, for the first five years of my faculty appointment, my salary primarily derived—more than 80 percent per year on average—from paying myself off of my own research grants. The vast majority of my salary support has derived from NIH grants, although some has derived from grants from pharmaceutical companies as well. Specifically, over the last three years, I have received research grants funded by the following companies: Gilead, Merck, Pfizer, Novartis, Astellas, and Enzon. Note that I have received little actual salary support from each of these industry research grants, as the money in the grants goes to pay for the supplies, equipment, and salary for the research assistants who help carry out the research.

Over the last five years I have also received lecture fees averaging under $10,000 per year combined from the pharmaceutical companies Pfizer, Astellas, and Merck. These lecture fees were for giving educational talks about infections caused by *S. aureus* and various fungal infections to doctors, pharmacists, and students.

I have also served on the Pfizer Visiting Professor grant-selection committee for the past two years. On that committee, my responsibilities are to peer review grants submitted from institutions that wish to have a professor from another campus come and visit their campus for a few days for educational purposes, paid for by the Pfizer grant program. As a grant-selection committee member, I do get compensated for my time spent reviewing grants and traveling to New York for one day, where the grant committee meets once per year. The compensation is well under $5,000 per year.

Over the last three years, I also have consulted for several pharmaceutical and biotechnology companies, including Merck, Arpida, Advanced Life Sciences, Theravance, and Basilea. Again, collective consulting fees from all of these companies combined have averaged under $10,000 per year.

Finally, I own shares in NovaDigm Therapeutics Inc., which, as I mentioned previously, is a startup biotechnology company founded by my colleagues and I to help translate the vaccine we discovered in our laboratories into a clinically useful tool.

My conflicts of interest are quite typical for academic physician-scientists. Nowadays it is nearly impossible to maintain an active academic laboratory in the United States relying exclusively on NIH funds. Because of its budget constraints, the NIH is currently rejecting ~90 percent of all grant applications. Fantastic grants from superstar labs are getting rejected as a matter of routine. There are just too many scientists doing research compared to the amount of dollars available to the NIH.

The NIH's budget has been frozen for several years now, as the country has continued to pour its resources into the Iraq and Afghanistan wars. Combine the cost of these wars with the recent economic crisis, and resulting growing federal budget crunch, and it seems likely that the NIH's budget will remain frozen in the years to come. Because of inflation, the NIH's spending power has actually markedly decreased during the past several years. Indeed, this is the first time since the 1980s that the NIH budget has failed to keep pace with inflation.[1] In this environment, active researchers who rely exclusively on NIH funding are likely to find themselves out of work, with shut-down labs. Supplemental grants from industry are critical to the maintenance of an active academic research program, and most successful academic researchers are now forced to seek such alternative sources of funding.

The reality is that in order to survive in academic research, it is mandatory to raise the money that pays for salaries, supplies, and equipment. Academic institutions provide very little money to their researchers, and they are getting stingier, not more generous, especially as federal, state, and local governmental support dries up. That means we have to look to other sources—wherever necessary—to bring in money to pay the bills.

Keep these facts in mind as you continue reading the rest of this chapter. I believe the most important step to be taken to resolve conflicts of interest is disclosure, and that's what I have tried to do here. It is up to you to decide for yourself how much these revenue sources have influenced my beliefs, versus how much my beliefs have been molded by the experience of seeing patients die from untreatable infections.

ARE PHARMACEUTICAL COMPANIES AS BAD AS ARMS DEALERS?

Let's consider this topic as objectively as possible. Arms dealers supply weapons designed to kill people to people who intend to kill people. While we're at it, let's look at tobacco companies. Tobacco companies sell a carcinogenic product to people who don't mind increasing their own risk of cancer by using the product. As a side effect, some of those people using the product will accidentally expose other people to potential carcinogens (i.e., secondhand smoke). Along the same lines, companies that make alcoholic beverages sell a product that can either be enjoyed or can be harmful, depending on the amount taken and the activities of the imbiber (i.e., are you driving or perhaps captaining a giant oil tanker through reefs in Alaska while drunk?).

How do pharmaceutical companies compare? Pharmaceutical companies do not make weapons. Pharmaceutical companies discover and develop new drugs. These drugs can have a wide range of effects, from lifesaving medications, to medications that treat symptoms of disease. Of course, these drugs can have side effects. But if used as directed, these drugs are enormously more beneficial to society than they are harmful to society. So, does this sound like pharmaceutical companies are worse than arms dealers? Or tobacco companies? Or even companies that sell alcoholic beverages?

To illustrate the point, it is worth spending a few moments to review the background for the book-turned-movie *The Constant Gardener*. The book drew its primary inspiration for pharmaceutical corporate misbehavior from a very unfortunate—and well-publicized—incident that took place between a Canadian generic drug company and a dedicated and honorable physician-scientist named Nancy Olivieri. At the time, Dr. Olivieri was the director of "the largest clinic in North America" that cared for children with genetic blood disorders that cause the children to need frequent blood transfusions.[2] Frequent blood transfusions cause children to get overloaded with iron, which results in severe health consequences, including death. There was a need to develop a treatment that could remove the excess iron from the body. While one such treatment already existed, it could only be given by injection, and had to be given frequently by injection for the rest of the child's existence, which was not a very convenient or happy way to treat a child. So there was tremendous interest in developing a pill version of such a drug.

Dr. Olivieri discovered just such a potential treatment. Originally the drug had been "licensed to Ciba–Geigy (now Novartis) but was abandoned by the company in 1993 because of [toxicity in animals]."[3] Nevertheless, desperation drove further testing of the drug, and Dr. Olivieri was able to collaborate with a generic drug company in Canada to manufacture the drug so it could be tested in humans. Let me reiterate the central point here—Dr. Olivieri was collaborating wth a *generic* drug company. Ironically, the movie and the public hate pharmaceutical companies and, in contrast, praise generic drug companies for providing cheap copies of drugs to the public. But the real-life story that inspired *The Constant Gardener* involved a generic drug company, not a pharmaceutical company.

As part of her grant from the generic drug company to fund testing of the drug, Dr. Olivieri signed a confidentiality agreement that gave the generic drug company veto rights over publication of any resulting data—this was standard practice at the time, and Dr. Olivieri had no choice in the matter. Had she refused to sign the confidentiality agreement, the company would not have funded or provided the drug for the research.

At first the human studies of the new drug seemed to be favorable. But in a second clinical study, for which Dr. Olivieri did not sign a confidentiality agreement, there appeared to be possible indications of liver injury in patients receiving the medication. Dr. Olivieri intended to release the data, but she was threatened with legal action by the generic company because of the confidentiality agreement she had signed as part of the grant to complete the first clinical trial. Dr. Olivieri did not agree that the confidentiality agreement signed for the first study held any sway over the results of the second study. So she defied the generic company and published the data. As a result, the company sued her. She also became the subject of a public smear campaign. Finally, the university where she worked removed her from directorship of her clinic, and did so while it was in the process of negotiating a multimillion-dollar donation from the very generic drug company with which Olivieri was having the dispute![4] It took years of legal wrangling to settle the dispute and for Dr. Olivieri to restore her good name. It is not surprising that Dr. Olivieri wrote that the movie *The Constant Gardener* "compelled me to relive years of harassment, false accusations and legal threats."[5]

But, to reiterate, the company at the center of the Olivieri conflict was a generic drug company, not a pharmaceutical company. It should also be emphasized that there were no threats on Dr. Olivieri's life or other forms of intrigue. The movie took license in going beyond harassment, which, of course, is bad enough.

There is also a very important epitaph to this story. The major pharmaceutical company that had originally abandoned the drug in question because of toxicity in animals has subsequently successfully developed an advanced-generation drug that is highly effective at removing iron from the body and did not have the toxicity problems that the earlier drug had. That new drug, deferasirox (Exjade), rapidly got approved by the FDA for use in humans based on data proving that it was both safe and effective to treat iron overload.[6] Exjade has rapidly become a cornerstone of treatment of iron overload in children and adults because it is taken as an oral medication rather than requiring frequent injections. Our own research (funded in part by Novartis) has shown that the same drug can starve very dangerous fungal pathogens of iron, thereby helping to treat potentially lethal infections in laboratory mice and potentially even in humans.[7] So, ironically, while the company that really was at the center of the controversy that inspired *The Constant Gardener* was a generic drug company, the pharmaceutical company peripherally involved in the Dr. Olivieri story not only did nothing wrong, it has created a new drug that helps treat sick children and may save the lives of people with infections!

Did you ever stop to consider why it is that people get so upset at drug prices in the United States? It seems to me that the reason people get so upset at pharmaceutical companies over the cost of medications is that people really want to take those medications. After all, if people didn't want to take those medications, they wouldn't care how much the medications cost. And the reason people want to take those medications is that the medications treat peoples' diseases!

This is not to imply that pharmaceutical companies are saints that are in business for the public good. Absolutely not! Let's get real here. Pharmaceutical companies, like all for-profit ventures, are in it for the money. That is their raison d'etre, and they will, of course, act aggressively, via lobbyists or any other mechanism available to them, to protect their profits. Again, I neither defend nor criticize this fact, I merely state it as a fact. Pharmaceutical companies are no more or less moral than any other large corporation in America. And the reality is, the priority of making money is the way corporations operate in a capitalistic society.

We can all agree that the US healthcare system is dysfunctional and does not work well for many of our citizens, and that healthcare is outrageously expensive. Let's face it, I work at a county hospital that cares largely for uninsured patients, so I am more than somewhat familiar with the problems of the US healthcare system. But the fact is, the general belief in society that US citizens have a right to affordable, needed drugs (such as new antibiotics) refers to a *moral* right, not a legal or constitutional right. This difference is critically important when considering how outrageous an idea it is to create incentives to stimulate pharmaceutical companies to make antibiotics.

If the availability of a specific class of needed drugs (e.g., antibiotics) was a legal or constitutional imperative, it would certainly seem reasonable to create legislation that forced pharmaceutical companies to make those drugs even if it was not considered cost optimal by the pharmaceutical companies to do so. In contrast, if availability of a specific class of needed drugs is a moral imperative, but not a legal or constitutional imperative, then society had better learn to work with pharmaceutical companies to reach a common goal of creating such drugs. That is, if we can't *force* pharmaceutical companies to make the drugs we want, we better figure out ways to get them to *want* to make those drugs. If we want antibiotics, and we cannot get antibiotics without the active participation of pharmaceutical companies, we had better learn to work with pharmaceutical companies so we can get those badly needed new antibiotics.

The fact that pharmaceutical companies make so much money off the sales of drugs makes many people uncomfortable, or even angry, at the thought of creating legislative incentives to stimulate new antibiotic development. Indeed, some have blamed pharmaceutical company price gouging for driving up the overall costs of healthcare in the United States. But, as we shall see, things aren't necessarily so black and white concerning drug price gouging versus legitimate pharmaceutical profits. There are shades of gray to this issue that merit careful consideration.

WHAT IS THE CONTRIBUTION OF PHARMACEUTICAL PROFITS TO US HEALTHCARE COSTS?

The perception that drug company profits are what drive up the cost of healthcare in the United States is inaccurate. In fact, the total amount of money spent on prescription drugs every year accounts for about 10 percent of total US healthcare costs (fig. 8.1).[8] I mentioned earlier that pharmaceutical companies have a 14 percent profit margin, meaning that 14 percent of their sales revenue results in profit. If 10 percent of US healthcare costs go to pharmaceutical sales revenue, and 14 percent of pharmaceutical sales revenue results in pharmaceutical company profits, that means pharmaceutical company profits account for 1.4 percent of US healthcare costs (10 percent × 14 percent = 1.4 percent). Hence, if pharmaceutical companies sold their drugs at a price equivalent to cost (i.e., they made no profit on the drugs), healthcare costs in the United States would actually decrease by only ~1.4 percent.[9]

Furthermore, despite the cost of prescription drugs, it turns out that prescription of pharmaceutical agents to treat diseases is one of the most

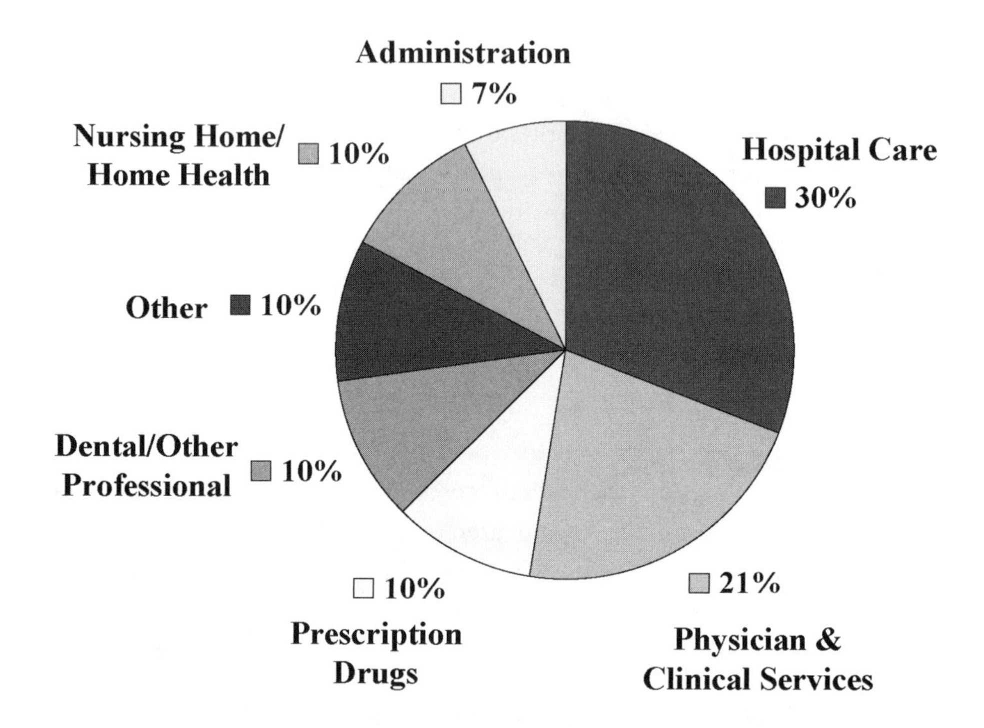

Fig. 8.1. US Healthcare Expenditures in 2004. (Reprinted with permission from http://www.healthguideusa.org.) Percentage of the overall US healthcare budget spent on various facets of healthcare. Only 10 percent of the overall US healthcare budget is spent on prescription drugs every year.

cost-effective tools a physician can use in all of modern medicine. The benchmark for a cost-effective intervention, by general governmental consensus, is $50,000 spent per year of life saved (adjusted for quality of life during that year—so-called Quality Adjusted Life Year, or QUALY).[10] Interventions that cost less than or equal to $50,000 per QUALY saved are generally accepted as cost effective. Table 8.1 shows where pharmaceuticals fit in the category of cost efficacy. It may surprise you that they rank fourth overall, and are even more cost effective than such widely accepted measures as screening tests for cancer (e.g., mammography or colonoscopy), widely accepted public-health measures (e.g., airbags or folic acid supplementation of food), and health education/counseling.[11]

TABLE 8.1. COST OF VARIOUS MEDICAL INTERVENTIONS

Intervention Type	Cost*
Vaccinations	$2,000
Delivery of Care in Appropriate Setting (e.g., ICU bed versus regular ward bed)	$6,000
Surgery	$10,000
Pharmaceutical Treatment	$11,000
Screening Tests (e.g., mammogram for breast cancer or colonoscopy for colon cancer)	$12,000
Other Public Health (e.g., car airbags, folic acid supplements in food)	$15,000
Health Education/Counseling	$20,000
Diagnostic Studies	$20,000
Medical Devices (e.g., pacemakers)	$40,000
Other Medical Procedures (e.g., blood transfusions, cardiac catheterization)	$140,000
OVERALL AVERAGE	**$12,000**

*Median cost (1998 dollars) per quality adjusted life-year saved

Adapted with permission from P. J. Neumann et al., "Are pharmaceuticals cost-effective? A review of the evidence," *Health Affairs (Millwood)* 19 (2000): 92–109.

So, healthcare costs in the United States are not driven by pharmaceutical costs, and in fact pharmaceutical expenditures are highly cost effective at prolonging life.

What then does drive healthcare costs in the United States? It turns out that a huge fraction of annual US healthcare expenditures occur in the last year, and even in the last several weeks, of patients' lives (fig. 8.2).[12] End-of-life expenses that drive up healthcare costs include enormous expenditures on elements of intensive care, including one-to-one nursing care; reimbursement for physician specialty care (including critical-care, surgery, pulmonary, infectious-diseases, cardiology, nephrology, and other specialists); laboratory tests (including costs of laboratory technician salaries, chemicals used in the tests, overhead on laboratory equipment,

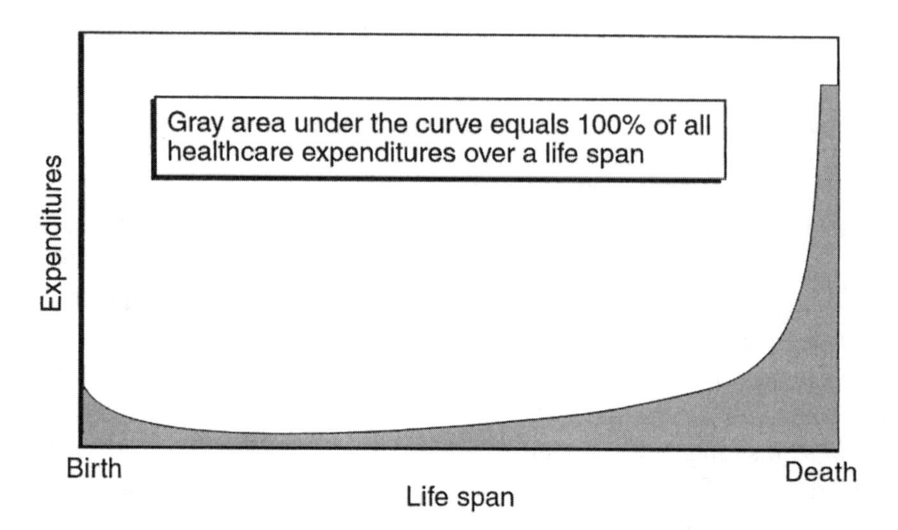

Fig. 8.2. US Healthcare Expenditures Over an Individual's Lifespan. (Reprinted with permission from http://www.medicaring.org/whitepaper/.) This figure shows the percentage of total healthcare expenditures incurred by an average American citizen (y axis) spread over an entire lifetime (x axis). The majority of healthcare dollars are spent in the last year, and even in the last months, of life.

etc.); reimbursement for respiratory therapists (who manage mechanical ventilators); the costs of advanced medical technology equipment, such as central venous catheters, digital monitoring equipment, hemodialysis, etc.; and the costs of hospice and/or skilled-nursing facility care. These and other elements of modern advanced medical technology incur hundreds of billions of dollars per year at the end of people's lives in order to prolong life to the maximum, even when doctors know that the patient is terminal and that all of the resulting care will, in the end, be futile. As Dr. Arnold Epstein, chairman of the Department of Health Policy and Management at the Harvard School of Public Health, recently said, "The U.S. is the one country in the world where [people] think death is optional."[13]

It is completely understandable, and expected, that people want to live as long as possible, and want their loved ones to live as long as possible as well. They tend not to think about what it costs society to help them stay alive, even if that prolongation of life is only a matter of weeks or a few months. But because healthcare is so expensive, the vast majority of terminal healthcare is covered by society, rather than by the individual patient's out-of-pocket costs (e.g., their annual deductible, or direct costs to the uninsured). Depending on the patient's financial and insurance status, such societal coverage may derive from an individual's health insurance policy (which is subsidized by everyone's premiums, not just the individual's), federal or state government programs like Medicare or Medicaid, or even local taxpayers who cover the costs of county healthcare for the uninsured who simply can't afford the out-of-pocket costs. So individuals who incur giant healthcare bills at the end of life are largely covered by society as a whole.

Who in society actually pays the healthcare bill? An insightful article was recently written in the medical journal *JAMA* that sought to answer this question.[14] The authors, Drs. Ezekiel Emanuel and Victor Fuchs of the NIH and Stanford, respectively, debunk the myth that government and employers pay most of the cost of healthcare in the United States. They point out that government pays for healthcare by taxing citizens. So the taxed citizens are actually footing the bill. When healthcare costs increase, government either increases taxes to foot the bill or they cut other parts of their budget, thereby reducing other services to the public. In either case, it is the taxpayer who really foots the bill for the healthcare. Furthermore, Emanuel and Fuchs point out that employer contributions to healthcare insurance are not typically taken out of the profits of companies but rather are accounted for by charging higher prices for goods and services and/or paying lower wages to employees.

Individual tax-paying citizens who work for a living are the ones bearing the burden of healthcare costs in the United States. So it is ironic that even those people who bitterly complain about the ridiculous cost of healthcare in the United States are likely to contribute to those costs by insisting on maximal care at the end of their and their loved ones' lives. Don't get me wrong, I am not at all implying that it is unreasonable to want everything to be done to prolong one's life. At the individual level it makes perfect sense. Nevertheless, the question still nags at the societal level. For example, is it unreasonable for an individual, terminally ill patient to want society to spend $300,000 on intensive-care unit expenses that will prolong that patient's life by a few weeks? If you are society, you might answer, "Yes, it is unreasonable." If you are that one patient, or his or her family, your answer will likely be very different.

The same intellectual construct can be directly applied to pharmaceutical costs. Let's say, for example, that you have congestive heart failure. Let's say that a group of medications have each been shown, independently, to prolong the life of people with congestive heart failure, and that the total cost of those drugs is $25,000 per year. You could also take alternative drugs, that are now generic, that don't prolong survival quite as well, and that are cheaper. Which would you, the patient, rather take? Do you want society to pay a huge price premium for a possible incremental benefit in the amount of time you get to live? The answer is almost certainly, yes.

Here's another example with which I am directly familiar. For half a century, a drug called amphotericin B deoxycholate has been a cornerstone of treatment for life-threatening fungal infections. The drug is very effective and very cheap ($5 per day, as an intravenous medication). But it is also toxic. It causes extremely uncomfortable shaking chills and severe muscle aches while it is being infused, so much so that patients who receive the drug have referred to it as "ampho-terrible." It also can cause kidney injury, which usually resolves if the drug is stopped at the first sign of trouble, but can in some cases be irreversible. Because of these toxicities, pharmaceutical companies have invented newer versions of amphotericin that are no more effective, but are much less toxic. Those new versions cost on average about $300 per day—sixtyfold more than the old drug. So, if you were my patient, with a life-threatening fungal infection, and I gave you the following two treatment options, which would you choose?

> Option 1: treatment with amphotericin B deoxycholate: your insurance will pay $5 per day ($70 for a two-week course) but you will have to tolerate very uncomfortable shaking chills and muscle aches on a daily basis, and take the risk of kidney injury that will probably get better if I monitor you closely, but may end up being permanent.

Or

> Option 2: treatment with a newer drug that is no more effective but will not cause toxicity and your insurance will have to pay $300 per day ($4,200 for a two-week course) for that drug.

So, which do you want? In fact, the choice has already been made over and over again. Even though they are no more effective, the newer amphotericin derivatives have virtually driven regular old amphotericin B from the market.[15] No one wants to take the old, unpleasant, toxic stuff when the new, more-pleasant, less-toxic stuff is available. No one really seems to focus much on the price difference. Is this the pharmaceutical companies' fault because they invented newer drugs that are less toxic and more expensive? Is it the physicians' fault because they are prescribing the more-expensive medications that are less toxic? Is it the patients' fault because they would much rather have the tolerable, less-toxic medications, and don't particularly care what it costs society to gain them that benefit?

You tell me.

Or, perhaps, instead of focusing on divvying up the blame, we should concentrate on understanding the complexity of the problem. The almost universal desire to prolong life regardless of cost is not surprising, and, as I said, it is quite reasonable from an individual's or his or her family's perspective. However, it is an example of the "Tragedy of the Commons." First described by Garrett Hardin in *Science* magazine in 1968,[16] the "Tragedy of the Commons" applies to scenarios where an individual acts to significantly benefit themselves, and as a consequence accepts as a tradeoff a small amount of overall harm to society at large. If only one person is so acting, the total harm to society is small. But when everyone in society undertakes that same action, the collective harm to everyone becomes enormous, and perhaps so much that it drives the system (i.e., our healthcare system) toward financial collapse. Given such a complex, societal problem, the expediency of assigning all of the blame to one group can only distract us from understanding the real causes of the problem—and prevent us from creating real solutions to the antibiotic crisis.

ARE PHARMACEUTICAL COMPANIES OVERCHARGING THE AMERICAN CONSUMER?

Regardless of the fact that pharmaceutical expenses account for only 10 percent of healthcare costs in the United States, there is significant anger and criticism that pharmaceutical companies fleece the American consumer by charging inflated prices for their drugs. The thinking goes, "Drugs are cheaper in Canada. Therefore, pharmaceutical companies are ripping off Americans and making United States healthcare unaffordable. Such companies should not be rewarded for their behavior with economic incentives."

But let me play devil's advocate for a moment and offer you a different perspective to consider. Please keep your mind open as you read the next few paragraphs. As you will see, I am not necessarily advocating one way of thinking over the other, I am merely trying to point out that the issue of pharmaceutical costs may be more complex than it initially appears.

An alternate point of view about the cost of drugs in the United States is that the increased prices paid by Americans for drugs is what actually enables pharmaceutical companies to succumb to political pressure in other countries and offer lower prices there. That is, for better or for worse, the American taxpayer may be subsidizing drug development for the rest of the world.

Everyone agrees that Americans pay more money for drugs than people in other countries. The perception is that people in other countries are paying the ideal price and Americans are paying prices that are too high. But, if Americans paid discount prices, like consumers in other countries, would pharmaceutical companies be able to continue to generate the capital necessary to invest billions of dollars of research money into creating the next generation of life-prolonging drugs? Or, in the face of shrinking profit margins, would they simply alter their business model and adopt a model more in tune with generic manufacturers, giving up innovation in return for more secure sales of drugs that are already on the market? Are American consumers, in fact, providing the capital that drives innovative, new drug discovery and development for the entire world?

Innovation is expensive and risky. Copying someone else's work is not. High profit margins enable an entrepreneurial business model, balancing costs of innovation and risk, ultimately enabling drug development. In fact, there is a very close relationship between annual pharmaceutical company sales and the amount of money these companies reinvest in research and development each year (fig. 8.3).[17] Over the last half century, during periods in which pharmaceutical sales have declined, so has reinvestment in research and development, and vice versa. From this "devil's advocate" perspective, the problem is that consumers all over the globe want to have their cake and eat it too. That is, they want pharmaceutical companies to continue to discover and develop lifesaving medications, but they don't want to pay pharmaceutical companies to do it.[18] Or, more specifically, global consumers are leaving it almost entirely up to the American consumer to subsidize new drug discovery and development.[19] Is the rest of the world taking advantage of the American consumer by gaining access to new technologies that Americans subsidize?

Keep in mind, price controls used in other countries to keep pharma-

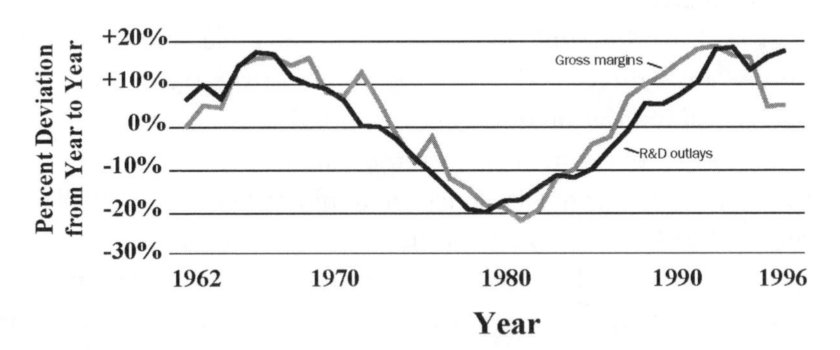

Fig. 8.3. Relationship between Annual Changes in Pharmaceutical Company Income ("Gross Margins") and Research and Development (R&D) Expenditures ("R&D Outlays"). (Reproduced with permission from F. M. Scherer, "The link between gross profitability and pharmaceutical R&D spending," *Health Affairs [Millwood]* 20 [2001]: 216–20.) This graph shows that, year by year, as pharmaceutical company income rises or falls, so do pharmaceutical company research and development expenditures. So, when companies make more profit, they spend more on research and development, fueling discovery of the next generation of medications.

ceutical costs low may have unintended side effects, particularly in suppressing innovative scientific research in those countries. Underscoring the potential negative impact of pharmaceutical price controls on research and development is the contrast between recent biomedical research and development activities in the United States and the European Union. In Europe, government price controls on prescription pharmaceuticals have become the norm. In the face of such price controls, over the last fifteen years there has been a complete reversal in the pharmaceutical research and development dollars spent in Europe versus in the United States, with the United States now far outstripping Europe (which used to far outstrip the United States).[20] By one estimate, US expenditures on biomedical scientific research now outstrip European Union expenditures by 60 percent.[21] As a result, US biotechnology companies account for more than 75 percent of global biotechnology revenues and global biotechnology research and development expenditures. Furthermore, the density of biotechnology patents per capita is 80 percent higher in the United States than in the European Union. Finally, there are approximately twice as many new drugs in development by pharmaceutical companies in the United States as compared to Europe.[22]

Can I say for sure that pharmaceutical price controls instituted in Europe are a direct and immediate cause of the astonishing reversal of biomedical innovation in Europe compared to the United States? No. But I can say the following: (1) when the European Union did not have pharmaceutical price controls, much more biomedical innovation was occurring there; (2) after pharmaceutical price controls were instituted in Europe, biomedical innovation declined there; (3) at the same time biomedical innovation has been exploding in the United States, where there are no pharmaceutical price controls; and (4) it is credible that price controls could stifle innovation in small biotechnology companies and could deflect larger companies from focusing development efforts in the region.

Indeed, from a broader perspective, data are certainly available to support the position that government regulations have been generally stifling European biomedical research and killing the biomedical scientific job market in Europe.[23] There has unquestionably been a major "brain drain" from Europe to the United States. Approximately 400,000 biomedical scientists who were educated and received their degrees in Europe now live and work in the United States, and of those surveyed, only 13 percent indicated they planned to eventually return to Europe.[24] These facts are compelling and merit careful consideration when evaluating the impact of pharmaceutical price controls on new drug discovery and development.

On the other hand, there are clearly dissenters who believe that government price controls on pharmaceuticals will not affect research and development outlays, and who are strong advocates of government-imposed price regulations on drugs.[25] To me, arguments as to why pharmaceutical price controls will not affect research and development seem naive. For example, among the arguments are that: (1) Europeans have successfully enacted price controls and look how well they are doing. But I've just shared with you data that suggest that price controls have had an extremely negative effect on European biotechnology research and development; (2) Future pharmaceutical research and development costs will be lower due to higher-tech, rational drug development. But as discussed in chapter 6, this widely claimed belief has never borne out, and in fact quite the opposite has been true. If anything, in the face of cutting-edge, rational drug design technology, fewer drugs are being developed now than previously and they are taking longer and costing much more money to be developed than previously; and (3) Pharmaceutical industry research and development risk is significantly mitigated by basic research funded by US taxpayers (e.g., NIH-funded research). But I've already explained in chapter 6 that the NIH does not fund drug development, and that pharmaceutical and biotechnology corporate research and development outlays are far greater than the annual NIH budget.

The negative impact of pharmaceutical price controls on innovative research and development has been emphasized over and over again.[26] This link is not merely pharmaceutical propaganda. Academic scientists and policy analysts have weighed in on the issue and concur that there is a clear link between pharmaceutical company revenues and research and development expenditures.[27]

The bottom line is that the issue of drug pricing is not nearly as cut and dried as many critics would like to make it. Pharmaceutical profits are important, to some extent, because they drive discovery of new, lifesaving medications. The consumer therefore benefits if drug companies are healthy, because new lifesaving medications then become available to the consumer. But drug costs also have to be balanced against the ability of consumers to pay. If consumers can't afford drugs, then there is no point in developing new ones. So there has to be a balance. What is the appropriate price balance for a given drug? I certainly have no answer. But then again, I don't think anyone else does either.

At the very least it seems safe to say that there is a complex relationship between drug prices, attempted controls on those prices, and research and development of new pharmaceuticals. The issue is not black and white, and the possibility of creating fiscally responsible legislative incentives to stimulate discovery of desperately needed new antibiotics should not reflexively be dismissed out of hand because of a visceral distaste of pharmaceutical companies. In the context of a critical need to develop new antibiotics for life-threatening, drug-resistant infections, surely a discussion about the merits and deficits of legislative incentives for new antibiotic discovery would be worthy of the attention of our politicians, and should be unfettered by bias or prejudice. Surely this debate should be resolved by focusing on the merits and deficits of the policy, and not on the basis of "toxic pharmaceutical politics."

IT'S TIME TO STOP THE FINGER POINTING

One of the biggest barriers to solving the crisis in antibiotic development is the ubiquitous problem of finger pointing. There are multiple parties involved in this complex problem, and each party wants to blame the others for the problem. As I've said, I have heard physicians, politicians, and laypeople express outrage at the idea of creating legislative incentives for pharmaceutical companies to develop new antibiotics. These people seem to believe that pharmaceutical companies have a societal responsibility to develop specific drugs that society needs, and that pharmaceutical companies should not be rewarded for abandoning this societal responsibility. In turn, pharmaceutical companies have blamed a significant component of the lack of antibiotic development on restrictive, burdensome, capricious, and murky regulatory requirements at the FDA, and on the lack of available guidance from the FDA on what constitutes acceptable clinical trial designs for testing antibiotics in development. As noted earlier, the NIH and other research organizations may be blamed for their own lack of government-sponsored drug development. Others might blame politicians for failing to recognize the significance of the problem and failing to act to address it. Finally, as mentioned, it is popular to blame physician misuse of antibiotics as the whole cause of the problem.

The central tenet under which the IDSA and its Antimicrobial Availability Task Force have operated is that lack of antibiotic development is a complex, interwoven societal problem that is *no one's fault*. For better or for worse, we live in a capitalistic society, and the for-profit motive is what drives business. Pharmaceutical companies do not have a constitutional or corporate responsibility to produce antibiotics. Rather, corporate directors have a legally codified fiduciary responsibility to invest their research and development dollars in a manner that maximizes the likelihood of return on investment. Indeed US corporations have been successfully sued by their shareholders for pursuing corporate policies that favor public good over corporate profits.[28] The laws of our society do not reflect the popular belief that pharmaceutical companies have a societal responsibility to develop antibiotics. Quite to the contrary, the laws of the land would put at considerable risk for a class-action shareholder lawsuit a company that chose to develop antibiotics to the detriment of its own bottom line. So it is completely unrealistic to expect pharmaceutical companies to be responsible for developing drugs that, while beneficial to the public good, do not maximize return on investment.

Keep in mind, large pharmaceutical companies have discovered, developed, manufactured, and brought to market nearly all of the antibiotics available today. Unfortunately, their motivation in this regard has deteriorated over time as more lucrative markets, including those for therapeutics to treat cancer, hypertension, hypercholesterolemia, arthritis and inflammatory diseases, and dementia, have arisen.[29] A critical void now exists, and it is incumbent upon society and our government—which holds the great responsibility of protecting the public's health—to step up to the plate with incentives to fill that void.

Furthermore, while some have criticized pharmaceutical companies for making "lifestyle" drugs in lieu of antibiotics, this belief is incorrect. Yes, lifestyle drugs for erectile dysfunction or bladder overactivity, and so on, have been brought to market. But those drugs reflect a small proportion of the drugs that pharmaceutical companies have brought to market over the last two decades. In lieu of producing antibiotics, pharmaceutical companies have been bringing to market critically needed drugs to treat cancer, high blood pressure, high cholesterol, diabetes, arthritis and inflammatory diseases, heart attacks, strokes, dementia, and more. Many of the readers of this book and their families are likely benefiting from such drugs right now, and many more will benefit from other revolutionary treatments discovered and developed in the future. So it is unfair—and worse, it is counterproductive—to lay the blame for lack of antibiotic development at the feet of the pharmaceutical companies.

Nor is it correct to blame the FDA, which has the incredibly challenging task of balancing the need to enable innovative products to come to market while still bearing responsibility for all safety issues related to such drugs. The mandate of the FDA is to protect public safety, not to stimulate badly needed drug discovery. The FDA has an absolutely enormous regulatory responsibility. They must ensure the safety of all drugs, medical devices, vaccines, high-tech biological therapies (such as antibodies), and foods that are marketed in the United States. They have a miniscule annual budget (less than $2 billion!)[30] with which to complete this Herculean task. It is frankly amazing that the FDA works as well as it does, and that drug approval is as efficient as it is, given the paltry resources available to the agency to complete its myriad tasks.

Nor is it correct to blame the NIH or other research-funding organizations, which create the basic science foundations upon which drug development occurs but have no mandate or capacity to develop drugs internally.

Nor is it correct to blame politicians. We live in an era in which the news is dominated by war, terrorism, and economic recession. When science or infectious diseases receive press, it is because of newly emerging infections that create mass hysteria, such as SARS, or "bird flu," or "swine flu," or bioterrorism (such as a potential smallpox attack), and rarely due to issues as banal as drug-resistant bacteria. Politicians answer to their constituents, and their constituents have not yet been convinced that problems of drug-resistant bacteria and lack of antibiotic development are as important or urgent as "bird flu," "swine flu," anthrax, or smallpox.

Finally, this problem is not the fault of physicians, who are increasingly faced with the impossible challenge of treating infections caused by organisms resistant to every antibiotic on the planet. To reiterate, this is a highly complex societal problem, and no one group is to blame. It is time to stop the finger pointing and to draw together to create solutions.

I have shared with you some of my own experiences with patients who contracted life-threatening, drug-resistant infections. Of course, the small sampling I described pales in comparison to the enormous number of such infections that occur every year in the United States and throughout the world. I think that if we were to have asked Mr. A., or Mrs. B., or Mrs. C., or Mr. D., or Mr. E., or Mr. F., or Mr. G., or any of the patients whose tragic stories are told on the IDSA's Web site[31]—or indeed any other patients who have contracted a drug-resistant infection—they probably would have been upset that each of the parties involved in the antibiotic crisis has spent so much time and effort trying to pass the blame on, pointing fingers at one another instead of joining forces to try to solve the problem.

It should be possible to create workable solutions to the antibiotic crisis without incurring significant additional societal healthcare costs. In the next chapter, I will propose such solutions. But if such solutions are to come to pass, we will have to draw together as a society and decide to stop playing the blame game.

CHAPTER 9

so what will work?

We have discussed the ideas that won't work. Now it's time to consider what can work, and we hope will work. The IDSA has spent years carefully considering strategies that could reverse the decline of antibiotic development. In its 2004 white paper "Bad Bugs, No Drugs: As Antibiotic Discovery Stagnates, a Health Crisis Brews," the IDSA proposed meaningful legislation that could effect change.[1] The debate has developed and matured over the years since the IDSA released its white paper, and recently my colleagues and I summarized the current thinking about effective legislative options that could stimulate discovery and development of antibiotics.[2]

PUSH AND PULL LEGISLATIVE INCENTIVES ARE NEEDED TO STIMULATE ANTIBIOTIC RESEARCH AND DEVELOPMENT

The first thing that must occur to lay the groundwork for developing new antibiotics is to come to a consensus on what the most problematic infections are that we are facing currently and that we will face in the future. To this end, it is necessary to establish a federal commission of experts that can create a list of priority pathogens that cause drug-resistant infections. Creation of such a commission is necessary for any legislative solution and must be the first step, because government has to have a prioritized list of pathogens that should be targeted by subsequent legislative incentives.

We have suggested referring to antibiotics that would be eligible for financial incentives as "priority antibiotics," and we have suggested defining priority antibiotics as those that treat serious or life-threatening infections caused by microbes that are resistant to available antibiotics.[3] Once a commission is in place and has created such a list of priority infections, the FDA would be able to rely upon the expertise of the commission to help it identify "priority antibiotics" as the drugs are being developed. Companies developing or receiving FDA approval for a newly developed "priority antibiotic," or a vaccine or other countermeasure that targeted prioritized infections, would become eligible for financial incentives. The commission would also be responsible for updating the priority infection list on a regular basis, so that incentives could be redeployed as needed based on discovery of new antibiotics and emergence of new antibiotic-resistant infections.

Once the federal commission has created a list of priority pathogens, it will be necessary to have legislative incentives available to apply to new priority antibiotics, either during the experimental/development phase of drug testing or after approval of the drug for use in humans. There are two fundamental ways to create incentives to stimulate research and development in antibiotics: *push strategies* and *pull strategies*. Push strategies are designed to give a boost to early phase products that need to cross the ever-widening chasm between basic science and early development milestones (for example, completion of manufacturing and early phase clinical trials). Pull strategies are designed to work at the opposite end of the spectrum, by providing financial rewards to companies that have already completed the developmental process and received FDA approval for use in humans of a new drug. Having both push and pull incentives is the optimal way to encourage new development of antibiotics. However, there are some significant, fundamental differences between push and pull strategies that affect how they must be designed, which companies will benefit from them, and how big they must be to be effective. Hence, the implementation of these strategies must be well thought out and coordinated.

In general, push strategies can involve much smaller dollar amounts than pull strategies because the costs of drug development at the preclinical and early clinical stage are much smaller than at late clinical stages. For example, if the block in developing an antibiotic is the inability of a small company or academic scientist to pay $2 million for manufacturing, or another $2 million for a phase I clinical trial, a relatively small push incentive could help get that product moving to the next phase. In contrast, for larger companies the block in developing an antibiotic is the need to invest $100 million in one or more risky phase III clinical trials. Obviously, therefore, the incentive to pull the antibiotic through the late stage, phase III clinical trials is going to have to be much larger.

Because push strategies can be smaller, their typical format is as a grant or contract offered by the federal government and awarded directly to the scientist or company that wants to develop a specific product. In contrast, because pull incentives must be several orders of magnitude larger than push incentives (hundreds of millions to billions of dollars, rather than millions of dollars), grants and contracts really cannot be used as pull incentives. Rather, pull incentives must come in the form of economic incentives related to the sale of a product, such as extended patents, periods of market exclusivity, and possibly major tax credits.

Note that such pull incentives need not—and given current economic conditions, perhaps should not—be in the form of direct pay subsidies, where companies are being handed cash by the government. In the aftermath of the recent economic crisis, the public's patience has worn thin on taxpayer bailouts and direct subsidies. But pull incentives can, for example, simply enable a company that is already selling an on-patent drug to sell that drug for a little longer before the patent expires. If judiciously applied, extended patents can be used to reward corporate behavior that is aligned with public health needs without requiring that the taxpayers dig deeper into their pockets.

Because of the nature of push and pull strategies, push strategies are more likely to be attractive to smaller companies and pull strategies are more likely to be attractive to larger companies. Smaller companies have little capital to pay for early milestones, and their business model tends to be to complete drug development into phase II clinical trials. Then they either partner with or get bought out by a larger company that can: (1) pay for phase III trials; (2) upscale the manufacturing to make enough drug to sell on the market; and (3) use an established distribution network and large marketing/sales force to promote and sell the drug once it is FDA approved, none of which small companies can do. Furthermore, smaller companies are likely to be interested in antibiotics and other drugs that sell in the tens to several hundreds of millions of dollars per year once developed, whereas very large companies tend not to be interested in such drugs.

In contrast to smaller companies, larger companies already have the capital to pay for early milestones and are therefore unlikely to be interested in push incentives. Also, as I mentioned in chapter 7, large pharmaceutical companies operate on an entrepreneurial business model that tends to not be amenable to government contracts, grants, and guaranteed markets. They would rather take the extra risk up front in order to secure a larger downstream payoff. Finally, larger companies tend to be interested in drugs that have a much larger market size, in the multiple hundreds of millions to billion dollars in sales per year. If a drug does not sell in the multiple hundreds of millions of dollars per year, its sales revenue will be insufficient to impact a large pharmaceutical company's bottom line. Unfortunately, for reasons elaborated in chapter 6, most antibiotics do not have large market sizes. Therefore, the pull incentives for antibiotics have to be substantial if they are going to stimulate large pharmaceutical company interest in developing antibiotics.

It is critical that these fundamental principles be understood if we are to create an effective array of incentives for antibiotic development.

SPECIFIC LEGISLATIVE INCENTIVES THAT COULD WORK

The easiest specific legislative incentive to establish, based on expansion of a preexisting program rather than creation of a new program, is to modify the FDA Orphan Drug Program to help encourage antibiotic development. So-called *orphan drugs* are those that are developed for markets that are too small to be of interest to drug companies. Specifically, the FDA defines an orphan drug as a "product that treats a rare disease affecting fewer than 200,000 Americans."[4] Orphan drugs are developed under the Orphan Drug Act, which is intended to provide incentives to companies that develop treatments for such rare diseases. In the early 1980s, Congress recognized that development of treatments for rare diseases is generally hampered by the small markets such diseases represent. That is, rare diseases provide insufficient return-on-investment to spur drug companies to develop treatments for such diseases—just as we've discussed is the case for antibiotics. Congress created the FDA Orphan Drug Act to mitigate this problem.

The Orphan Drug Act has multiple provisions to encourage drug development for diseases with small market sizes, including both push and pull incentives. For example, companies developing orphan drugs are eligible for tax incentives and actual grant funding from the FDA to defray the costs of the clinical research they carry out in support of the orphan drug (push incentives). In addition, once a company has successfully developed an orphan drug, and receives FDA approval to market that drug, the company gets a seven-year period of marketing exclusivity for their new treatment (a potential pull incentive). This marketing exclusivity means other companies cannot market competing drugs for seven years. Of course, companies have to apply to the FDA for orphan drug status for their drug, and the FDA carefully reviews such applications to ensure that the drug being developed actually meets the criteria as defined in the law. Only those treatments that receive an official designation of "orphan drug" from the FDA become eligible for these incentives.

By all accounts, the Orphan Drug Act has been an extremely successful piece of legislation. In the twenty-five years since the Orphan Drug Act passed, over two hundred orphan drugs and biological products have been brought to market using its mechanisms.[5] Given the small market sizes represented by the diseases those orphan drugs treat, it is very likely that few or none of those treatments would have been brought to market without the Orphan Drug Act. Given the success of this act, one idea regarding antibiotic development would be to expand orphan drug criteria to include antibiotics that target prioritized infectious pathogens. Appropriate financial push and pull incentives for new antibiotic development could include grants to encourage clinical development of these products (push), tax credits for research and development costs (push), a period of market exclusivity (pull), and federally funded advance-purchase commitments or other "promised markets" (pull).

Such an expansion of the scope of the Orphan Drug Program may well encourage biotechnology and small pharmaceutical companies to expand their presence in the antibiotic market. However, the effect of such a program on large pharmaceutical companies would likely be minimal. As mentioned, large pharmaceutical companies are looking for billion-dollar blockbuster drugs, which few antibiotics are. So, even with tax credits or grants to defray development costs, the size of the pull (i.e., the downstream market size) is going to be insufficient to cause many large companies to become interested in antibiotic development. Furthermore, as mentioned already, pharmaceutical companies are leery of accepting government money in the form of grants or guaranteed markets, as all government money comes with government strings attached.

Another possible legislative solution to the antibiotic crisis is to include antibiotics in the definition of "countermeasures" found in the newly enacted Biomedical Advanced Research and Development Authority (BARDA). BARDA was signed into law in 2006 with bipartisan congressional support. The law as it currently stands creates a single entity within the Department of Health and Human Services that is responsible for research and development of medical countermeasures to bioterrorism and natural disease outbreaks. BARDA is modeled after the highly successful Defense Department program, Defense Advanced Research Projects Agency (DARPA), which has been responsible for the development of numerous cutting-edge defense industry technologies, such as stealth technology. Like DARPA, BARDA is intended to foster collaborations between academic scientists and commercial technology experts to more rapidly translate cutting-edge science into cutting-edge technology.

Similar to use of the Orphan Drug Program, expansion of BARDA's scope to include financially supporting new drugs, vaccines, or other interventions that target priority pathogens would help stimulate antibiotic development among smaller biotechnology companies by enhancing academic-biotechnology collaborations. However, also as for the Orphan Drug Program, the potential effect of expansion of BARDA's scope on large pharmaceutical company research and development is less clear.

It has also been pointed out to me that there is tremendous fear in the antibiotic development world about rare but serious toxicities that can occur from antibiotics, which create an enormous potential liability risk after a drug is approved for sale in the United States. The problem is that during the clinical trial period, the number of patients exposed to experimental antibiotics is typically in the hundreds to perhaps a thousand or so, at most. Unfortunately, sometimes drugs can have rare side effects, which affect one in fifty thousand or so patients. Even if these side effects are serious (e.g., liver injury), they likely will not have occurred during the clinical trial process, because too few patients will have been exposed to the drug during the clinical trials for these rare toxicities to manifest. As a result, there have been several examples over the last decade of very useful, very effective antibiotics receiving approval for sale in the United States that subsequently were found to cause rare but serious toxicities, after the drugs had already been approved by the FDA for widespread commercialization.

The problem is, when these rare side effects of antibiotics are subsequently discovered, it is not the FDA that is liable for damages, but rather the companies that developed the drugs. This fear of getting sued over a toxicity that is too rare to be detected during clinical testing, and manifests only after the drug is approved for commercialization, is a significant disincentive for antibiotic development. One potential way to remove this barrier to developing new antibiotics is to create a communal insurance policy for postmarketing antibiotic injury, modeled after the Vaccine Injury Compensation Program (VICP). The VICP provides precisely this type of medical-legal protection. The protection is actually provided to the patient, as payments are made directly to patients who have sustained injury that is possibly related (indeed in many cases it is not even certain that such injuries are actually caused by the vaccines in question) to very rare adverse effects from vaccines approved for use in the United States. The critical features of the VICP are that it is a trust fund, managed by the US Department of the Treasury, not by pharmaceutical companies, and that its capital derives from a very small excise tax ($0.75) charged against the sale of each dose of vaccine in the United States. So, it is functionally an insurance policy that provides monetary awards to rare patients who may have sustained an injury from a vaccine, and it is funded from the sale of the vaccines themselves, rather than by direct US government support.

The VICP was created because of the universal agreement in the medical community that widespread vaccination is a required feature of public health, and that companies should not be discouraged from developing and manufacturing those vaccines due to fear of being sued from very rare adverse effects, including those that may have had nothing to do with the vaccine an injured patient received. Since the availability of active antibiotics to treat drug-resistant bacteria is of a similar public-health imperative, it is logical to consider creation of a similar type of trust fund to provide insurance for very rare side effects possibly related to newly created antibiotics targeting priority pathogens.

A theoretical concern that could be raised about instituting a federal liability policy for rare antibiotic side effects is that removal of liability risk could encourage pharmaceutical companies to be less vigilant for evidence of side effects during development of a new antibiotic, since they would not be financially liable for such side effects after a drug was approved by the FDA for sale in the United States. This concern could be mitigated by provisions in the federal liability policy that would exclude from coverage side effects common enough to have been detected during clinical trials. Indeed, a specific allowable limit of the frequency of side effects could be set such that if the frequency of the side effect exceeded the limit, the side effect would not be covered by the federal program. For example, if a side effect occurred more commonly than one in every twenty-five thousand patients exposed to the drug, it would not be covered by the liability program. The onus would be on the pharmaceutical company to prove what the actual frequency of the side effect was in order to seek coverage from the liability insurance program.

Setting a limit to the frequency of a side effect for it to be covered by a federal antibiotic liability program would have three significant advantages. First, it would limit the overall risk to the liability program—the program likely wouldn't be able to remain solvent if it had to pay out coverage for side effects that are common enough to be detected during clinical trials. Second, it would mitigate the concern that the liability program would enable pharmaceutical companies to be less vigilant about detecting side effects during clinical trials, because the company would have to have adequate data to prove that the side effect had definitely not occurred during clinical trials if it wanted coverage from the liability program. Just saying "We didn't look for it" wouldn't be enough. And third, it would actually encourage pharmaceutical companies to be more vigilant about detecting side effects "postmarketing"—that is, *after* the FDA had approved the new antibiotics for sale in the United States. The enhanced vigilance for side effects after FDA approval would result from the fact that if the pharmaceutical company could not specifically prove that the side effect was occurring less frequently than the policy cutoff, they would not get liability coverage for that side effect. If the companies were not actively monitoring for rare side effects, they likely would not be able to prove that the frequency of those side effects was indeed less than the policy cut-off. In contrast, currently there is no incentive for companies to monitor for side effects after drug approval. And since there is no active surveillance for side effects after a drug is approved for sale in the United States, rare side effects may be discovered late, after hundreds of thousands or even millions of patients have been exposed to the drug. Encouragement of pharmaceutical companies to conduct active surveillance for rare side effects could enable far earlier detection of the side effects, limiting the overall number of patients exposed to the drug before the side effects became recognized. Hence, a properly designed federal liability program could actually improve public safety.

The other major category of legislative financial incentives to stimulate antibiotic development is extension of patent life (a pull strategy). For example, companies producing a priority antibiotic could apply to the FDA for extended patent protection, which would improve the return on investment of such antibiotics. Note that, as proposed, the patent extension would be from six months to two years, at the FDA's discretion. So we are still talking about a limited patent extension. While controversial, such a concept has been mentioned previously in a US Government Accountability Office (GAO) report,[6] and critics of the idea tend to lack a complete understanding of precisely what has been proposed.[7] Again, direct patent extension would likely stimulate interest in antibiotic development among smaller biotechnology companies. However, as I mentioned before, the annual sales of most antibiotics are in the hundred-million-dollar range, well under the multibillion-dollar blockbuster range. While such sales volume is alluring to smaller companies, some large pharmaceutical companies may be less motivated to sell such drugs for a longer period of time. Hence, the potential impact of extending patent life on antibiotics on large pharmaceutical participation in drug development is unclear.

Finally, we have come to the lightning rod of possible legislative incentives: transferable patent extensions, also known as "wild-card" patent extensions. Of the potential solutions I have mentioned, wild-card patent extension provides by far the largest pull incentive and is generally acknowledged to be, by far, the incentive most likely to successfully stimulate new antibiotic development by large pharmaceutical companies.[8] Under this concept, any company receiving FDA approval for a new antibiotic that treats a priority pathogen could be granted a patent extension of six months to two years on one other drug that company sells. Since, as noted earlier, annual sales of individual antibiotics very rarely reach the magical billion-dollar mark, large pharmaceutical companies are unlikely to be motivated to develop new antibiotics if the incentive is prolongation of sales of the antibiotic itself. In contrast, pharmaceutical companies are very interested in prolonging on-patent sales of their billion-dollar blockbuster drugs. Hence, allowing pharmaceutical companies to transfer a patent extension to a drug of their choice within their portfolio would be a powerful incentive for them to successfully develop new antibiotics.

The concept of wild-card patent extension has been extremely controversial, and it merits a more extensive discussion. In fact, even though it is by far the most likely incentive to stimulate new antibiotic development by large pharmaceutical companies, wild-card patent extension has been so controversial that it is no longer considered feasible and has been taken totally off the table due to the perception that it could incur a significant financial cost to society. The IDSA is no longer publicly advocating this legislative solution because it is so unpopular on Capitol Hill.

The major concern about wild-card patent extension has been that extending patents on billion-dollar blockbuster drugs will drive up the cost of healthcare in the United States by prolonging the time before such drugs are available in cheaper, generic forms. Transferable patent extension has even been characterized as a "boondoggle" for the pharmaceutical industry.[9] It is perhaps not surprising that the most vocal critics of the wild-card patent extension program are generic drug manufacturers and their lobbyists. Generic drug manufacturers perceive that extension of patents, even if for only six months to two years as proposed, will delay their ability to begin profiting from the sales of generic copies of those drugs. However, generic-manufacturer criticism of patent extensions likely will prove to be shortsighted; if pharmaceutical companies do not discover, develop, and seek regulatory approval for new antibiotics, generic manufacturers will have no new antibiotics to copy as generics in the future, even as sales of old generic antibiotics drop precipitously due to rising antibiotic resistance. Furthermore, from a public-health perspective, making cheaper generic versions of already existing drugs does not address the problem of rising drug resistance and the increasing incidence of pan-resistant, lethal infections; only innovative discovery of new antibiotics can address this problem. Generic manufacturers are incapable of such innovative drug discovery.

The controversy over wild-card patent extension strikes an emotional core in both sides of the debate. Probably because of this visceral, emotional reaction, two critical factors in this debate have been overlooked. First, wild-card patent extensions will not *increase* healthcare costs. The patent extensions would be applied to drugs already on patents. Extending those patents would prolong the sales of branded drugs. That is, the current cost of those drugs would be continued for an extra six months to two years. So, healthcare costs derived from those drugs would not be increased—they would be continued at the same cost for a slightly longer period of time before declining.

Second, and more important, a critical concept that has been almost totally unappreciated in this controversy is the potential for newly developed priority antibiotics to mitigate the dramatic costs imposed on society by antimicrobial resistance. In fact, antibiotic-resistant infections cost the US healthcare system tens of billions of dollars annually.[10] Drug-resistant infections cost tremendously more to treat than nonresistant infections because they cause extra medical complications requiring additional medical treatments and longer hospitalizations than do nonresistant infections. The availability of new antibiotics has the potential to markedly reduce the cost of these multi-drug-resistant infections.

My colleagues and I recently sought to analyze this exact dilemma and to determine if the potential for new priority antibiotics to decrease the societal costs of resistant infections would more than account for the continued cost to society of wild-card patent extension.[11] We evaluated a hypothetical scenario in which a single new priority antibiotic was developed that had activity against infections caused by the multi-drug-resistant bacterium *Pseudomonas*. To be conservative, we assumed that the new antibiotic would not be effective against any other drug-resistant pathogen (when, in reality, any antibiotic effective against drug-resistant *Pseudomonas* would very likely be effective against other resistant bacteria, and would therefore reduce the cost of infections caused by these other bacteria as well). We also assumed the transferred patent extension would be for two years, when the actual proposal is a range of six months to two years depending on the FDA's perceived benefit of the new antibiotic. To be extra conservative, we also charged the annual sales of the new antibiotic against the overall cost-benefit of the program. Finally, we assumed that only 50 percent of the excess costs of drug-resistant infections would be recoverable by having a new antibiotic with which to treat the infections.

With these conservative assumptions, we found that a single use of wild-card patent extension would prevent society from saving approximately $7.7 billion over the first two years (due to the continued sales of the on-patent drug, which otherwise would have gone generic), and would cost an additional $3.9 billion over the following eighteen years (due to the sales of the new antibiotic). On the other hand, we found that the new antibiotic would reduce the healthcare costs of multi-drug-resistant *Pseudomonas* infections by approximately $1.35 billion per year. In balance, therefore, the program would become cost neutral by ten years after approval of the new antibiotic, and by twenty years, society would save a net total of $4.6 billion.[12] So, in our analysis, wild-card patent extension was not only cost effective, it was cost saving! Based on this analysis, it is time for the experts to reengage in the debate about the merits of wild-card patent extension as a tool to promote antibiotic development.

Another concept that could make wild-card patent extension more acceptable is the proposal by the IDSA that companies profiting from extended patents would be required to earmark a certain percentage (e.g., 10–20 percent) of profit from the extended patent toward further antibiotic research and development. Hence, the public would further benefit from additional drug discovery resulting from the use of wild-card patent extension. I'll go a step further. I would propose that the reinvestment of profits into research should be funneled through the National Institute of Allergy and Infectious Diseases (NIAID), which, as mentioned previously, is the branch of the NIH that deals with infectious diseases. Specifically, companies would be required to donate to a special fund at NIAID 10 to 20 percent of the profits they made from use of wild-card patent extension. NIAID would then fund competitive research grants from academic investigators studying basic mechanisms of antibiotic resistance. In this way, companies would be encouraged to develop new drugs, patients would have new drugs to treat their antibiotic-resistant infections, and scientists would have further research dollars to lay a greater groundwork for future antibiotic development by companies—an "upward spiral" of success.

It is essential that an open, honest, dispassionate debate take place about wild-card patent extension and other incentives for antibiotic development. Visceral dislike of pharmaceutical companies has no place in this debate. The issue here is, how badly do we want new antibiotics to treat life-threatening infections? Society and the government must make an informed decision, taking into account all factors, and must then come up with a plan to stimulate new antibiotic discovery. If incentives are rejected, then it is up to the rejecters to come up with a viable alternative plan. To date, no such viable alternative plan has been forwarded.

VACCINES AND OTHER IMMUNE SYSTEM–ENHANCING STRATEGIES ARE COMPLEMENTARY, NOT ALTERNATIVES

Since defects in host immune systems predispose patients to developing drug-resistant infections, finding ways to boost the immune system should be an effective way to prevent or treat such infections. There are scientists all over the country working on such strategies, including my colleagues and me. Aside from basic public-health measures, such as clean drinking water and sanitation, vaccination is by far the most effective and the most cost-effective way to reduce the global burden of infectious diseases. Vaccines and other immune-enhancing strategies have tremendous potential to reduce the overall burden of infections and infection-related deaths, and should be a major focus of both government and industrial research and development.

However, as a practicing infectious-diseases specialist, I need to emphasize that it is naive to believe that immunological strategies will be able to completely eliminate the need for new antibiotics. Vaccinated patients do get infections. No immune-enhancing strategy is ever going to be 100 percent effective. Also, vaccines can't be developed against every possible pathogen, and when the burden of a vaccine-targeted disease is reduced by vaccination, other diseases begin to appear that were unimportant before. The bottom line is, antibiotics will always be required to treat people who have active infections. Vaccines and antibiotics complement one another, and both are needed.

CHAPTER 10
what can you do to help?

PARALLELS: GLOBAL WARMING AND ANTIBIOTIC DEVELOPMENT

Al Gore taught me a lesson. Mr. Gore struggled for three decades with a problem of a similar nature to the declining availability of new antibiotics. He fought against overwhelming resistance to convince people that global warming was a real threat. Politicians hated the idea, because they believed that acting to fix global warming would have significant, negative economic implications. For many years, the public didn't know anything about the problem. And when the public began to learn about the problem, they primarily received information from businesses, lobbyists, and politicians that was in conflict with the actual data being generated by the scientists, who had no voice to explain reality to the public. As a result, paralysis set in and nothing happened for many years, even as the critical problem grew worse and worse.

Whether or not you are a political fan of Mr. Gore, and whether or not you liked *An Inconvenient Truth*, the book or the documentary, it is clear that finally, after so many years of trying, Mr. Gore seems to have turned the corner on the issue of global warming. While the scientific community has been essentially unanimous about global warming for many years, political action on the issue has been impossible due to recalcitrant politicians. Once again, politicians' attitudes tend to reflect the attitudes of the constituents that put them into office. For many years the general public remained unconvinced about global warming, so there was no pressure on politicians to act to solve it. But now, as the public has finally come around to understanding and believing in global warming, even the most obstinate of critics, George W. Bush, has been forced to admit that global warming is a real phenomenon.[1]

How did Mr. Gore do it? How did he turn the tide? Can we learn something from his methods that we can apply toward solving the antibiotic crisis? It was while I was watching his documentary film, *An Inconvenient Truth*, that I gained a new perspective on the problem we are up against in trying to fix the antibiotic problem. We had been attacking the problem by lobbying Congress to act. But what I now believe is really needed to make Congress act is to attack the problem at the other end, by creating a grassroots movement among the voting public. Once a grassroots movement has begun, it will inherently bring pressure to bear on political leaders. Again, politicians work on issues that are important to their constituents. If a specific issue is not recognized by the public to be of critical importance, it is unlikely to be a priority for politicians. So we need to make society aware of the scale of the problems of antibiotic resistance and lack of antibiotic development. Until we do, legislative solutions are not going to be forthcoming.

How big is the hurdle before us? Not only has the general public not known about the lack of antibiotic development, or the extent of the problem, it turns out that many physicians do not know either. Nor do general medical journals appear to be particularly interested in the problem. Several leading medical journals have declined even to consider publishing materials related to this issue. There seems to be a belief that this issue is important only to infectious-diseases specialists, not to physicians in general, or to the public. This belief baffles me. So I can understand Mr. Gore's surprise, expressed in *An Inconvenient Truth*, that for so many years his colleagues in Congress seemed to lack an interest in global warming. I feel the same surprise about the lack of urgency prevalent among members of the medical community about the problem of antibiotic development.

As infectious-diseases specialists and the IDSA try to sound the general alert and fire off warning klaxons about this imminent, dire problem, what we are finding is that some of those in charge of pressing the alert buttons are yawning and rolling their eyes. But even a cursory glance at figure 5.1 (p. 108) demonstrates unequivocally that there has been a dramatic decline in antibiotic development over the last twenty-five years. That's the key message that, for some reason, is not getting through.

Many eye rollers and nit pickers work in comfortable ivory towers surrounded by paperwork and don't see patients for a living. Many are not physicians. They're not down in the trenches, awash in rotting flesh, trying to stem the juggernaut tide of virulent, resistant microbes. They've never had to tell a patient's family that their loved one is going to die of an infection because society has run out of antibiotics to use against it. Hence, they feel no sense of urgency about the problem. The data in figure 5.1 leave no question about the scope of the problem. There is no nit-picking one's way out of the obvious conclusion that emerges from the data in that graph: we are not getting new antibiotics that we desperately need so we can treat drug-resistant infections.

This is where you come in. You've been armed with knowledge of the problem. Now, I ask you to help us spread the word. Help us fire the warning signal into the air. We need a grassroots movement to catch the eyes of the politicians. Make this issue—the need to develop new antibiotics—a priority in your local, state, and federal voting. If enough people are talking about this issue, and people begin to vote on it and to demand solutions to it, politicians will begin to act on it.

In the previous chapter I listed numerous feasible legislative solutions that could be effective at turning around the decline in antibiotic development. These solutions are not mysterious, and it did not take a Manhattan Project of subatomic physicists, rocket scientists, or Mensa members to come up with them. These solutions are obvious, and their proposal is the result of innumerable discussions among the IDSA, legal counsel, pharmaceutical and biotechnology representatives, and congressional staffers. We *know* how to fix the problem, society just lacks the *will* to do it. Help us build that will.

Finally, you may not agree with the solutions I listed in the last chapter. You may disagree with the rationale I laid out justifying legislative incentives. You may think that I've overlooked other potential solutions to the problem. That's fine too. I don't claim to have a monopoly on answers, and neither does the IDSA, as it emphasized in its 2004 "Bad Bugs, No Drugs" white paper.[2] But rather than nit picking or eye rolling at the solutions thus far proposed, if you disagree with the conclusions or suggestions I and others have made, then suggest alternative solutions.

The most important thing is to create a public debate. If others have better solutions that are more palatable to politicians or the public, fine—let's discuss and act on them. But we must do something.

PUBLIC DEBATE

A good example of a reasonable exchange of ideas between two sides that recognize the problem but disagree on the solution is reflected in an exchange with opposing views published in the medical journal *Lancet Infectious Diseases*. The original editorial that started the exchange was written by Mr. Kevin Outterson and colleagues in 2007.[3] In response to their paper, I subsequently wrote a letter to the editor,[4] in collaboration with Dr. Henry Masur, who was at the time the president of the IDSA. The IDSA endorsed the letter we put together. Our response spurred Mr. Outterson[5] to write a rebuttal. Both our letter and Mr. Outterson's response were published in the same issue of *Lancet Infectious Diseases* in 2008.

Mr. Outterson is not a physician or a scientist. He is a lawyer at the Boston University School of Law. In the original editorial piece, he and his colleagues argued that the real problem facing the developed world was inappropriate use of antibiotics rather than lack of antibiotic development.[6] Of course, I've spent the prior chapters explaining why I profoundly disagree with that belief. Eliminating inappropriate antibiotic use is important to help buy time for us to create new antibiotics to deal with resistant infections, but it is not a long-term solution to the problem of antibiotic-resistant infections.

In their original editorial piece, Outterson and his colleagues wrote many things with which I disagree. Nevertheless, I need to emphasize that while we disagree on cause and effect, as you will see I actually like some of Mr. Outterson's proposed solutions to the problems of antibiotic development. So, as it turns out, this exchange of ideas is a microcosm of what can work in this debate. Furthermore, as I emphasized in my letter in response to the original editorial piece, Mr. Outterson and his colleagues should be congratulated for actually offering counterproposals, rather than sitting there and rolling their eyes without bothering to suggest alternatives, as others have done.

I would like to go through some of the debate between Mr. Outterson and myself because the issues raised in this debate highlight some critical concepts that must be understood if we are to create real solutions to the antibiotic crisis. In their original editorial, the following are examples of assertions with which I disagree:[7]

> Wildcard patents provide an incentive for companies to deliver more antimicrobial drugs, but the costs may be staggering. A 2-year wildcard patent extension on the top ten selling drugs would protect more than $125.3 billion in global annual sales from generic competition. The global cost of granting just ten wildcard patent extensions will likely exceed $40 billion, more than $4 billion per new drug (table). If the wildcard proposal allows stacking of multiple extensions on a single blockbuster drug, then the cost might double.

I described in chapter 9 why I disagree with this thinking about wild-card patent extension, and why wild-card patent extension could be beneficial. First of all, patent extension does not increase the costs of healthcare. It merely continues the current costs of an on-patent drug for a slightly longer period of time before the drug goes generic and becomes less expensive. That is, patent extension continues costs, it does not increase costs. Second, two years is the outer limit of what has been previously proposed for wild-card patent extension. The proposal has been for a period of six months to two years, with the duration for each newly developed antibiotic based upon a rational cost-benefit analysis for that drug. Third, wild-card patent extension would only apply to antibiotics that treat "priority" pathogens resistant to current antibiotics. Given the difficulty in developing new antibiotics, not to mention new antibiotics that treat priority/resistant pathogens, it is extraordinarily unlikely that ten such drugs would be eligible for the wild-card patent extension at any one time. It might take decades for ten such drugs to get developed.

Fourth, the issue of "stacking" of multiple patent extensions confuses me. If a single drug company were to develop three priority antibiotics eligible for wild-card patent extension, it should make no difference if they used the patent extension three times on the same drug or three times on three different drugs. The cost to society is the same.

Finally, and most importantly, as discussed in chapter 9, we need to take into consideration the cost savings resulting from having a new antibiotic with which to treat resistant infections. I have explained that our own analysis of the wild-card patent extension idea suggested that society would actually *save* money from the program because the costs of multi-drug-resistant infections would be decreased by having a new antibiotic with which to treat the infections.[8]

In my letter to the editor in response to the Outterson editorial, I explained this exact point.[9] In our analysis, we found that wild-card patent extension would prevent savings of about $7.7 billion over the first two years by extending the patent on a drug and keeping it from going generic, and cost about $3.9 billion over the next eighteen years due to the sales of the newly developed antibiotic.[10] In contrast, we found that the savings from the program due to having a new antibiotic with which to treat drug-resistant infections would be $2.7 billion over the first two years and $13.5 billion over the next eighteen years. Therefore, over a twenty-year period, wild-card patent extension would actually result in a net savings to society of $4.6 billion. In essence, we found that the program would not only pay for itself, in the long run it would decrease healthcare costs in the United States!

Yet, when Mr. Outterson responded to my letter, he wrote: "In a recent article, Spellberg and colleagues estimated the US cost of a single 2-year wildcard patent at $7.7 billion. Our methodologies differ substantially, and yet both numbers greatly exceed estimated costs of drug development."[11] Unfortunately, this statement mentions the cost portion of our calculation, but fails to mention the savings portion of our calculation resulting from diminished cost of drug-resistant infections. To reiterate, we found that wild-card patent extension *saves* money in the long run by reducing the cost to society of drug-resistant infections.

> A major problem with conservation is reimbursement: virtually no one pays for antimicrobial conservation. In the USA the opposite occurs: hospitals are reimbursed for nosocomial infections. This is a travesty of the highest order. Many proven techniques could be encouraged through changes in reimbursement. . . . Healthcare providers should receive substantial financial rewards for achieving benchmarks in infection control and the appropriate use of antimicrobial drugs.

The above statement reflects a sentiment that I commonly hear expressed, which is that we don't need new drugs, we just need to stop "tolerating" the occurrence of drug-resistant infections in hospitals. The problem is that available scientific data do not support the concept that antibiotic resistance occurs because of lack of physician or hospital reimbursement for antimicrobial conservation. I have explained, in detail, the causes of antibiotic resistance in prior chapters.

The belief that hospital-acquired infections reflect bad medicine is becoming pervasive, as is the belief that it is an easy fix to put pressure on doctors to get rid of those infections. Unfortunately, there are no data to support the idea that it is within our power to eliminate hospital-acquired infections by practicing better medicine. Can we do better at reducing them? Yes, I'm sure we can. Should we try to? Yes, absolutely we should try. Our ongoing, important, best efforts to continue to improve antibiotic stewardship and infection control will hopefully lead to a reduction in the natural-selection pressure that continues to increase the frequency of hospital-acquired, drug-resistant infections. But our best efforts will never lead to the complete elimination of hospital-acquired infections and therefore the elimination of the need for new antibiotics.

As a well-publicized example, let's look at the MRSA epidemic in hospitals. There are very powerful patient advocacy groups and political forces that have developed the belief that if we test every person admitted to the hospital for the presence of MRSA, and then put all of those patients in isolation if they are carrying MRSA, we could decrease hospital-acquired infections. Politicians have even passed legislation mandating such activities in certain states. But what does the actual scientific data on these practices tell us?

A comprehensive review of the practice of active surveillance for MRSA on patients admitted to hospitals found that there had been a total of twenty studies done to investigate if it works to decrease MRSA infections.[12] However, of those twenty studies, "None . . . were of good quality." Therefore, "definitive recommendations [about whether or not to use active surveillance for MRSA in hospitals] cannot be made."

There have been two more recent, large-scale studies evaluating whether or not universal screening for the presence of MRSA on patients admitted to the ICU or to the hospital works. The first study found that this practice didn't work at all.[13] In contrast, the second study did find a benefit of checking everyone on admission to the hospital for MRSA.[14] However, even in this second study, MRSA infections were by no means eliminated during the intervention period. They were statistically reduced but still occurred at high rates. One of the world's leading experts on MRSA infections, Dr. Chip Chambers, has pointed out that at San Francisco General Hospital (where he is chief of infectious diseases), they did not actively screen for MRSA on patients and did not isolate those patients discovered to have MRSA infections (until a new California state law was passed based on inadequate scientific evidence requiring that they do so). Nevertheless, their rate of hospital-acquired MRSA infections in the blood was still tenfold lower than the incidence of infection achieved in the second, successful study even after introduction of active surveillance and decolonization strategies![15] So, does active screening for MRSA in hospitals have any benefit at all? We simply don't know. If a hospital that doesn't do it has a tenfold lower incidence of hospital-acquired infection than a hospital that does, how can we justify mandating that all hospitals screen for MRSA?

It is also critical to realize that much of the emphasis being placed on screening for MRSA infections is being driven by for-profit companies that are trying to sell rapid detection diagnostic kits, as pointed out by Drs. Edmond and Eickhoff in their fascinating editorial "Who is steering the ship? External influences on infection control programs."[16] The authors point out that a recent lecture series promoting MRSA screening was sponsored by the maker of a test to screen for MRSA. That lecture series passed through several cities, used speakers who were specifically proscreening, without also using speakers who were not convinced by the evidence, and also had a session "in which company spokespersons addressed the audience."

Finally, it is critical to emphasize that isolating patients who are infected with antibiotic-resistant bacteria does *not* guarantee that the bacteria won't spread to other sites. I've already told you the story of Mr. A., who brought drug-resistant *Acinetobacter* into my hospital. Mr. A. was isolated very promptly. His bacterium escaped anyway. Similar experiences have been described in hospitals all over the world, over and over again.[17] One major reason for this is that bacteria can live on inanimate objects and in the environment, not just on people. Bacteria can be spread by contact with inanimate objects and environments in the hospital, even if the person who brought the bacterium into the hospital is no longer there.[18]

Finally, there is another "cost" of isolating patients who are found to be colonized with but not infected by MRSA or other drug-resistant bacteria. Patients who are placed in isolation may actually be harmed by the placement in isolation. Because they are in isolation, an additional barrier is created for healthcare professionals, including doctors, nurses, phlebotomists, respiratory therapists, and so on, to visit those patients. In fact, as Drs. Edmond, Eickhoff, and Diekema have pointed out, studies have shown that placing patients in isolation reduced by 50 percent the number of visits those patients receive from healthcare professionals, and this reduction in visits resulted in much less frequent monitoring of vital signs, higher rates of depression and anxiety, and increased risk of physical ailments such as bedsores and falls.[19]

We just don't know how effectively we can reduce hospital-acquired infections. It is true that we may be able to slow down the rate of such infections with better infection-control practice and more stringent adherence to hand washing. But hospital-acquired infections will always occur. And we will always need antibiotics with which to treat them.

> If we want more agents that are effective against pathogens, the most direct and efficient route may be to greatly increase long-term government grants from the US National Institutes of Health and their counterparts throughout the Organisation for Economic Co-operation and Development.

Again, the idea that government scientists will do our drug development for us is a common counterargument to the need to stimulate pharmaceutical company development of antibiotics. But I've already explained that the idea that the NIH will develop antibiotics reflects a lack of understanding of what the NIH does. You can increase grant funding all you want, but if all you do is increase NIH funding, and there is no private/corporate participation, you'll end up with a tremendous fund of basic knowledge but no drugs resulting from the knowledge. As I've discussed in prior chapters, there is a fundamental difference between science and technology. Excluding private/corporate participation from the process will result in a dearth of technology development, regardless of what happens with the fundamental science.

I certainly agree that the NIH budget should be expanded as much as is feasible, so that our scientists can continue to lay the groundwork for future technology development. But it is not enough to expand basic science research. We must also continue to expand NIH's nascent programs that are designed to fund the translation period of research, bridging the basic science research to the partnership with private capital that is necessary for technology to be developed. We also need to create a federally funded network of clinical investigators to help design and conduct complicated clinical trials for antibiotic-resistant infections. Furthermore, in partnership with expanding these NIH programs, we need to encourage venture capitalists, biotechnology companies, and pharmaceutical companies to continue to invest research and development dollars in developing drugs that are discovered by NIH-funded scientists.

Although I profoundly disagree with Mr. Outterson on many of his views on the antibiotic crisis, I believe that the debate we continue to have is not only healthy but essential. Furthermore, I agree with him on many of his proposed solutions to the problem.

Mr. Outterson concludes his rebuttal letter by asking, "Could we achieve better medical results at a lower price through a combination of approaches, including vastly expanded National Institutes of Health and European grants, multibillion-dollar prizes, changes to reimbursement, investments in diagnostic and vaccine innovation, supporting the prudent use of antimicrobials, and improving global surveillance infrastructure? I share Spellberg's conviction that an open and transparent debate of these issues is of the first importance."

While I strongly believe that patent extension, either direct or transferable, is the most powerful method to encourage new antibiotic development, all of the other suggestions Mr. Outterson makes are also valuable and insightful options. As I've said, I do agree that expanding grant funding to scientists will help propel drug discovery forward by laying a firmer and more extensive groundwork for future drug discovery—as long as corporations are there to continue the actual drug development. Just as private/corporate participation is necessary to translate such basic science into drugs, private/corporate partners absolutely require that a groundwork of basic science be laid so they can translate the science into technology. This is a yin-yang relationship, and both scientists and technologists are needed. As I mentioned, it is also necessary to expand the translational research portfolio at the NIH, to bridge the scientists and the technologists.

Another interesting concept is Mr. Outterson's suggestion of exploring the potential to use "prizes" to encourage drug development. He suggests an alternative, powerful "pull" incentive that I did not discuss in the prior chapter. He writes, "Indeed, why not offer a . . . prize to the first effective treatment for each of the five high-priority pathogens identified."[20] A "prize" would be an interesting pull incentive, and could be explored. However, it would represent a true subsidy and therefore might be less palatable to the taxpaying public than patent extension.

As I've already indicated, I agree that we do need to support antibiotic stewardship, public-health infrastructures to track drug-resistant infections, and better diagnostic and vaccine technologies to make early diagnoses and to prevent drug-resistant infections.

Finally, and most importantly, Mr. Outterson and I are on exactly the same wavelength when it comes to the open and transparent debate of this process. In fact, I believe we've begun the process of a public, open, and transparent debate. It is time for others to join in, to change the debate from a one-on-one format to a chorus of voices calling for change.

WE NEED YOUR STORIES

If you or anyone you know has been affected by an antibiotic-resistant infection, you can also help to put a "human face" on the problem. I urge you to visit the IDSA Web site and to submit your story at http://www.idsociety.org/staaract.htm. The IDSA needs to collect such stories to underscore the problem to politicians. At http://www.idsociety.org/badbugsnodrugs.html you can also learn more about the IDSA's Bad Bugs, No Drugs advocacy campaign, and you can contact your members of Congress to urge them to act.

PROTECT YOURSELVES FROM ANTIBIOTIC-RESISTANT INFECTIONS

Finally, there are positive steps you can take to help protect yourself and others from antibiotic-resistant infections. First and foremost, help us preserve antibiotics by not asking for them if you have a cold. Don't expect your physician to treat a viral infection with an antibiotic. Understand the critical need not to use antibiotics inappropriately, so we can buy ourselves time to build the will to restimulate antibiotic development.

Second, the key to reducing your risk of acquiring an infection in the community is good hygiene. Sounds pretty basic, I know, but that's the straight dope. It may not be expensive or high tech, but it helps. Frequent hand washing is the single most important step you can take. It's not 100 percent effective, but it will reduce your risk of transmitting microbes to others, and it will reduce your risk of having microbes transmitted by others to you. Wash your hands with soap and water, or an alcohol-based hand gel or solution, on a regular basis. Shower or bathe daily with soap and water. Be sure to wash your entire body with soap when you shower, as drug-resistant bacteria like MRSA can live in a variety of places and seem particularly fond of the armpits, groin, and buttocks. Wash linens and other items shared in the household regularly. Good personal hygiene and simple cleanliness in the home, at school, and at work are your most effective weapons to protect yourself from acquiring infections in the community.

The single most important thing you can do to protect yourself from hospital-acquired infections is, keep yourself out of the hospital! No duh, right? I know it sounds obvious, but if it is so obvious, why does our society continue to be increasingly inundated by medical problems like obesity, diabetes, heart disease, and others, which put people in hospitals and could be effectively diminished by lifestyle changes? If you stay fit, eat right, exercise, stop smoking, and comply with basic safety recommendations (e.g., wear seat belts, wear helmets while biking, drive safely, etc.), you will decrease your risk of developing medical problems that lead to hospitalization. I'm certainly not implying that you can completely avoid health problems by living a healthy lifestyle, but you can reduce your risk of developing them.

If you do have to go into the hospital, work with your physicians, nurses, physical therapists, and other ancillary staff to shorten as much as possible your inpatient stay and transition yourself to home care as soon as you can. Again, this may sound simple and obvious, but you'd be surprised how common it is to hear patients say they are reluctant to be "rushed" out of the hospital. Physicians cringe when they hear this. There is an old saying among physicians who take care of hospitalized patients: "The longer they stay, the longer they stay." That is, every day a patient is in the hospital increases his or her risk of developing a complication of hospitalization—such as a hospital-acquired infection—which will then further prolong his or her hospitalization. We see it over and over again.

Finally, help scientists secure the resources that are needed to better understand antibiotic-resistant infections. As mentioned previously, under the Bush administration, the budget of the National Institutes of Health actually decreased relative to inflation for the first time in several decades.[21] When you go to the polls to vote, make funding of public health and scientific research a priority. Help us secure the resources needed to lay the groundwork for real prevention and treatment strategies for antibiotic-resistant infections.

CHAPTER 11

consequences and conclusions

There is no question that antibiotic development has dramatically declined over the last twenty-five years. Unfortunately, this crisis in new antibiotic availability is occurring at exactly the same time that the world is experiencing an explosion in the frequency of multi-drug-resistant infectious diseases for which we require new antibiotics. Furthermore, there is little to no public appreciation of this convergence of events. The causes of the problem are complex, but the potential solutions are relatively straightforward. The real barrier to execution of the solutions is a lack of political will, which is largely due to the lack of public appreciation of the problem.

So, what happens if no solution is forthcoming? How bad could it really get?

The consequences of the failure to create new antibiotics could be catastrophic. We live in a time in history when we expect to cure infections. Antibiotics really are miracle drugs. There is no other field of internal medicine in which physicians routinely cure their patients. Rather, physicians working in most fields of medicine expect to control symptoms of infection and to prevent progression of disease. For example, if you have diabetes, the best we can do is give you insulin or insulin-stimulating drugs to try to keep your blood sugar under control. But we can't cure the diabetes or fix your pancreas so you don't need those diabetic drugs anymore. If you have high blood pressure, we can give you pills that lower your blood pressure, which will prevent longer-term complications. But we can't cure your high blood pressure so you can stop taking those medications that are keeping your blood pressure low. If you have arthritis, we can give you pills to treat the pain and inflammation, but we can't make the underlying arthritis go away. We can cure some cancers, but others are cured very rarely or almost never. The examples go on and on. An expectation of cure is virtually unique to the field of infectious diseases, and indeed it is a major draw to the field among young physicians seeking career paths.

This period of history, in which we have so viewed infectious diseases, has lasted beyond the duration of a single career. It has been nearly seventy-five years since sulfa drugs first became available, and nearly seventy years since manufactured penicillin was first used to treat an unfortunate police officer with a staph skin infection. Since that time, several generations of physicians have come and gone. To put this in perspective, senior physicians that remain in practice today trained in the 1960s—that is, already twenty-five to thirty-five years into the antibiotic era. There are a very few even-more-senior docs around today who are still active and who trained in the mid to late 1950s, twenty years into the antibiotic era. But there really are no active physicians around who have ever practiced medicine without antibiotics. This lack of firsthand experience with the preantibiotic era has the effect of making the medical community and society at large myopic and complacent, taking antibiotics for granted and believing that antibiotics are simply a part of our world and always will be.

But let's look at the history of antibiotics from another perspective. After all, the first patient treated with penicillin received the drug in 1940. That is only sixty-nine years ago, less than the average life span of people in countries with advanced medical technology. That means there are millions of people alive today who have lived in a world without antibiotics. We have been aware of antibiotics for no more than 0.002 percent of the four-million-year history of our species. It is a tribute to the absolutely astonishing efficacy of antibiotics that in a period less than a human lifetime, we have come to take them for granted so much.

The best description I've seen of the desperation that imbued every aspect of medicine before antibiotics came along was in a wonderful book called *The Youngest Science: Notes of a Medicine-Watcher*.[1] Recently my colleague from the IDSA's Antimicrobial Availability Task Force, Dr. W. Michael Scheld, referred me to the book, and it was truly eye opening. The author, Dr. Lewis Thomas, wrote the book late in his career, when he was chancellor of the Memorial Sloan-Kettering Cancer Center in New York. *The Youngest Science* is a retrospective look at his life and career in medicine.

Although it was published a quarter-century ago, the book has a unique perspective that is highly relevant to the antibiotic-resistance problems of the twenty-first century. Dr. Thomas grew up in the preantibiotic era, and what's more, his father was a small-town physician who trained at the turn of the twentieth century and practiced through the 1930s. As a child, young Lewis would often go on house calls with his father. During these house calls, Lewis saw how medicine was practiced before sulfonamides. Furthermore, Lewis graduated from medical school and began his internship training in 1937, the first full year in which sulfonamides were widely available in the United States. So, he saw the before and after of antibiotics in medicine, and that gave him a very unique perspective.

Dr. Thomas wrote that during his father's career, the only "really indispensable drug in the whole pharmacopoeia" was morphine, to treat pain.[2] When young Lewis would go on house calls with his father, his father taught him about the ethos of medicine. "I was not to have the idea that he [Dr. Thomas senior] could do anything much to change the course of [his patients'] illnesses," Dr. Thomas junior wrote. "It was important to my father that I understand this; it was a central feature of the profession, and a doctor should not only be prepared for it but be even more prepared to be honest with himself about it."

Dr. Thomas further described the prescriptions that physicians wrote in that era:

> These were fantastic formulations . . . each one requiring careful measuring and weighing by the druggist. The contents were a deep mystery, and intended to be a mystery. The prescriptions were always written in Latin, to heighten the mystery. The purpose of this kind of therapy was

> essentially reassurance. A skilled, experienced physician might have dozens of different formulations in his memory, ready for writing out in flawless details at a moment's notice. . . . They were placebos, and they had been the principal mainstay of medicine, the sole technology, for so long a time—millennia—that they had the incantatory power of religious ritual.[3]

Ultimately, such prescriptions "gave the patient something to do while the illness, whatever, was working its way through its appointed course."

Dr. Thomas's description of "tonics" was particularly amusing. He wrote that tonics "were good for bucking up the spirits; these contained the headiest concentrations of alcohol. Opium had been the prime ingredient in the prescriptions of the nineteenth century, edited out when it was realized that great numbers of elderly people, especially 'nervous' women, were sitting in their rocking chairs, addicted beyond recall."[4] The most popular tonic at Boston City Hospital during the 1930s, where Lewis Thomas worked as a Harvard medical student, was "Elixir of I, Q, and S," which consisted of tiny amounts of iron, quinine, and strychnine (!) dissolved in what Dr. Thomas described as "the equivalent of bourbon."

Medical education in the era immediately before sulfonamides became available was dominated by the curriculum promulgated by the legendary physician Dr. William Osler. The medical education that Dr. Thomas received at Harvard taught him to focus on making a specific diagnosis, so that an accurate prognosis could be provided to the patient and/or the patient's family.[5] Students were explicitly taught to avoid having an expectation of changing that prognosis. That is, discussion of treatment was specifically avoided. The Oslerian revolution was a backlash against nineteenth-century medical chicanery, in which doctors made up treatments galore that were almost always kooky, sometimes violent, and not infrequently downright harmful to patients. Dr. Thomas listed the following examples, "bleeding, cupping, violent purging, the raising of blisters by vesicant ointments, the immersion of the body in either ice water or intolerably hot water," etc.

Of his internship, Dr. Thomas wrote:

> For most of the infectious diseases on the wards of Boston City Hospital in 1937, there was nothing that could be done beyond bed rest and good nursing care.
>
> Then came the explosive news of sulfanilamide, and the start of the real revolution in medicine.
>
> I remember the astonishment when the first cases of pneumococcal and streptococcal septicemia were treated in Boston in 1937. The phenomenon was almost beyond belief. Here were moribund patients, who would surely have died without treatment, improving . . . within a matter of hours . . . and feeling entirely well within the next day. . . . We became convinced, overnight, that nothing lay beyond reach for the future. Medicine was off and running.[6]

Indeed. In the twenty-first century physicians *expect* to cure infections. We feel we've failed when we can't cure them. That's how far the pendulum has swung away from the Oslerian belief that physicians should admit to themselves that they cannot alter the course of their patients' diseases, which dominated medicine before sulfonamides just three-quarters of a century ago.

But just because we are in an "antibiotic era" now does not mean that we will always be. People tend to project their own experiences forward in history. We have airplanes now, so we will have even more advanced aircraft in the future. We have computers now, which undoubtedly will be far more powerful in the future. We have firearms now, and will probably have even more advanced weapons, such as lasers, in the future. Once a technology is established in society, there is little reason to expect that it will be lost in the future.

But antibiotics are unique. There is no other technology that becomes less effective the more it is used. In this regard, antibiotics may be thought of as being akin to limited precious resources, such as oil, or forests, or wetlands. And it is entirely conceivable that we will reach a time when our precious antibiotic resource is exhausted.

The most obvious consequence of failure to replenish the antibiotic resource is a marked increase in deaths and disability from untreatable infections. Without effective antibiotics, it is very likely that infectious diseases will increase as a leading cause of death throughout the world. The good news is that fewer people would die of heart disease and cancer. The bad news is that the decline in heart disease and cancer deaths would be caused by people dying of infections before they were old enough to develop heart disease or cancer.

But the consequences of lack of antibiotic development go far beyond an increase in untreatable infections. The very availability of effective antibiotics has revolutionized public health and has been responsible for enabling countless advancements in all aspects of medical care. Antibiotics have been critical to the development of advances in surgery. Complicated and/or prolonged surgery is feasible because antibiotic prophylaxis markedly reduces the incidence of infection from surgery. Antibiotics also allow surgery of infected tissues or of tissues that are at particularly high risk for infection (such as abdominal surgeries, in which intestines are manipulated). Similarly, chemotherapy for cancer was developed only in the context of available antibiotics. Without effective antibiotics, so many people would die of infections after having their immune systems wiped out by chemotherapy that most cancer treatments would be impractical. Similarly, transplantation of hearts, lungs, livers, kidneys, pancreases, and bone marrow (or stem cells) would be impossible without effective antibiotics.

Effective antibiotics have also been critical to advanced medical treatment of patients with trauma, burns, and battlefield injuries. The battlefield death rate from infection plummeted in World War II, in large part because of the availability of the miracle drug, penicillin. As resistance caught up with penicillin, the standard antibiotics used by the military to prevent or treat wound infections have become more powerful. The same thing is true of victims of civilian traumatic injuries, such as car accidents and industrial accidents. But what happens when new, powerful antibiotics no longer become available?

As we've discussed, cancer chemotherapy wipes out patients' immune systems, leaving them defenseless against common bacteria or fungal infections. The global and US populations are aging. Cancer rates are going to continue to increase. Cancer chemotherapy is going to continue to become more and more aggressive, as oncologists try harder and harder to cure recalcitrant cancers. Those chemotherapy regimens are going to cause even more severe impairments of cancer patients' immune systems. The fact is, if effective antibiotics are not available, chemotherapy will become impossible to administer. The cure would become worse than the disease, because antibiotic-resistant infections would simply kill the patients.

Finally, the very field of intensive-care medicine depends on the availability of effective antibiotics. Patients with heart attacks, strokes, traumatic injuries, and respiratory failure require intensive care with large plastic catheters in their veins, with nutrition being administered via the vein, and with artificial respirations provided by mechanical ventilators. The infection rates due to each of these interventions are extremely high. If we did not have effective antibiotics with which to treat these infections, these interventions would cause as many or more deaths as they would save.

Ironically, the very advances in medical care enabled by effective antibiotic therapies have, in turn, created enormous populations of increasingly immunocompromised patients who develop infections caused by increasingly resistant microbes that require treatment with newer, more powerful antibiotics. As mentioned, as global and US populations continue to age, this upwardly spiraling need for intensive care with catheters and ventilators, for increasingly aggressive cancer chemotherapy, and for heart, abdominal, and other complicated surgeries are all going to continue to increase. While we have come to take for granted such elements of modern medical care, their continued utility depends in large part on the continued availability of effective antimicrobial therapy.

We must not be fooled into believing that we are somehow entitled to effective antibiotics. We should not expect that they will always be available. Microbes will never stop adapting to selective pressure we impose upon them. For this reason, if we do nothing, we run the risk that some day, historians will look back with nostalgia upon a golden "antibiotic era," when doctors could treat infections with miracle drugs. In that bleak future, medicine would descend into a "postantibiotic era," and infectious diseases would once again reign supreme.

It is incumbent upon all of us to work together to solve this complex societal problem. We need to educate our friends and neighbors. We need to help the IDSA accumulate compelling patient stories to help stimulate action. We need to write to our congressional representatives. We need to make this issue a voting priority. The lives of our family, our friends, and others are in our hands. We must act now to protect our future.

endnotes

FOREWORD

1. S. Junger, *The Perfect Storm* (New York: W. W. Norton, 2007).

INTRODUCTION: TO DIE FROM AN UNTREATABLE INFECTION—NO ONE IS SAFE

1. After four years of medical school, students receive their Medical Degrees (MD), and become doctors. The new doctors then undergo *residency* training to specialize in their field of choice. *Internship* is the first year of residency. One of the fields of medicine a doctor can specialize in is internal medicine. An internal medicine residency lasts for three years (intern year, junior resident year, senior resident year), after which the resident becomes an internal medicine specialist. To subspecialize in a specific field of internal medicine (e.g., cardiology, pulmonary-critical care, nephrology, gastroenterology, rheumatology, endocrinology, infectious diseases), the physician must then undergo yet another training program, called a *fellowship*. An infectious-diseases fellowship typically lasts for two years. Thus in this book, a *resident* is a doctor in training to specialize in internal medicine and a *fellow* is a medically licensed specialist in internal medicine who is training to subspecialize in infectious diseases.

2. L. C. McDonald, "Trends in antimicrobial resistance in healthcare-

associated pathogens and effect on treatment," *Clinical Infectious Diseases* 42, Suppl. 2 (2006): S65–71; J. E. McGowan Jr., "Resistance in nonfermenting gram-negative bacteria: Multi-drug-resistance to the maximum," *American Journal of Medicine* 119 (2006): S29–36, discussion S62–70; D. L. Paterson, "The epidemiological profile of infections with multi-drug-resistant *Pseudomonas aeruginosa* and *Acinetobacter* species," *Clinical Infectious Diseases* 43, Suppl. 2 (2006): S43–48; P. R. Rhomberg et al., "Clonal occurrences of multi-drug-resistant gram-negative bacilli: Report from the Meropenem Yearly Susceptibility Test Information Collection Surveillance Program in the United States (2004)," *Diagnostic Microbiology and Infectious Disease* 54 (2006): 249–57; L. S. Munoz-Price and R. A. Weinstein, "*Acinetobacter* infection," *New England Journal of Medicine* 358 (2008): 1271–81.

3. Munoz-Price and Weinstein, "*Acinetobacter* infection"; N. E. Aronson et al., "In harm's way: Infections in deployed American military forces," *Clinical Infectious Diseases* 43 (2006): 1045–51; "*Acinetobacter baumannii* infections among patients at military medical facilities treating injured U.S. service members, 2002–2004," *MMWR Morbidity and Mortality Weekly Report* 53 (2004): 1063–66; K. A. Davis et al., "Multi-drug-resistant *Acinetobacter* extremity infections in soldiers," *Emerging Infectious Diseases* 11 (2005): 1218–24; J. S. Hawley et al., "Susceptibility of *Acinetobacter* strains isolated from deployed U.S. military personnel," *Antimicrobial Agents and Chemotherapy* 51 (2007): 376–78; A. Jones et al., "Importation of multi-drug-resistant *Acinetobacter* spp. infections with casualties from Iraq," *Lancet Infectious Diseases* 6 (2006): 317–18; H. C. Yun et al., "Bacteria recovered from patients admitted to a deployed U.S. military hospital in Baghdad, Iraq," *Military Medicine* 171 (2006): 821–25.

4. Munoz-Price and Weinstein, "*Acinetobacter* infection"; M. E. Falagas and E. A. Karveli, "The changing global epidemiology of *Acinetobacter baumannii* infections: A development with major public health implications," *Clinical Microbiology and Infection* 13 (2007): 117–19.

5. "Bad bugs, no drugs: As antibiotic discovery stagnates, a public health crisis brews," Infectious Diseases Society of America (IDSA), http://www.idsociety.org/pa/IDSA_paper4_final_web.pdf (accessed April 30, 2005); B. Spellberg et al., "Trends in antimicrobial drug development: Implications for the future," *Clinical Infectious Diseases* 38 (2004): 1279–86.

CHAPTER 2. INFECTIONS, ANTIBIOTICS, AND ANTIBIOTIC RESISTANCE

1. "The Office of the Public Health Service Historian: Frequently asked questions," http://lhncbc.nlm.nih.gov/apdb/phsHistory/faqs.html (accessed October 19, 2006).

2. Ibid.; B. Spellberg, "William H. Stewart, MD: Mistaken or maligned?" *Clinical Infectious Diseases* 47 (2008): 294.

3. A. S. Fauci, "Infectious diseases: Considerations for the 21st century," *Clinical Infectious Diseases* 32 (2001): 675–85; J. M. Hughes and F. C. Tenover, "Approaches to limiting emergence of antimicrobial resistance in bacteria in human populations," *Clinical Infectious Diseases* 24, Suppl. 1 (1997): S131–35.

4. M. Burnet, "Viruses," *Scientific American* 184 (1951): 51.

5. M. Burnet, *Natural history of infectious disease* (Cambridge: Cambridge University Press, 1962).

6. G. B. Pier, "On the greatly exaggerated reports of the death of infectious diseases," *Clinical Infectious Diseases* 47 (2008): 1113–14.

7. R. G. Petersdorf, "The doctors' dilemma," *New England Journal of Medicine* 299 (1978): 628–34.

8. R. G. Petersdorf, "Whither infectious diseases? Memories, manpower, and money," *Journal of Infectious Diseases* 153 (1986): 189–95.

9. "Deaths by cause, sex and mortality stratum in WHO Regions, estimates for 2002. World Health Report-2004. Annex Table 2," World Health Organization, http://www.who.int/whr/2004/en/09_annexes_en.pdf (accessed March 30, 2007).

10. R. W. Pinner et al., "Trends in infectious diseases mortality in the United States," *JAMA* 275 (1996): 189–93.

11. G. S. Martin et al., "The epidemiology of sepsis in the United States from 1979 through 2000," *New England Journal of Medicine* 348 (2003): 1546–54; D. C. Angus et al., "Epidemiology of severe sepsis in the United States: Analysis of incidence, outcome, and associated costs of care," *Critical Care Medicine* 29 (2001): 1303–10; D. C. Angus and R. S. Wax, "Epidemiology of sepsis: An update," *Critical Care Medicine* 29 (2001): S109–16; "National vital statistics reports: Deaths—leading causes for 2002," National Center for Health Statistics, http://www.cdc.gov/nchs/data/nvsr/nvsr53/nvsr53_17.pdf (accessed March 18, 2007).

12. R. M. Klevens, J. R. Edwards, C. L. Richards et al., "Estimating

healthcare-associated infections and deaths in U.S. hospitals, 2002," *Public Health Reports* 122 (2007): 162–66.

13. M. D. Williams et al., "Hospitalized cancer patients with severe sepsis: Analysis of incidence, mortality, and associated costs of care," *Critical Care* 8 (2004): R291–98.

14. J. Diamond, *Guns, germs, and steel: The fates of human societies* (New York: W. W. Norton, 1997).

15. G. Majno, *The healing hand: Man and wound in the ancient world* (Cambridge, MA: Harvard University Press, 1975).

16. S. C. Eisenbarth et al., "Crucial role for the Nalp3 inflammasome in the immunostimulatory properties of aluminium adjuvants," *Nature* 453 (2008): 1122–26.

17. G. W. Hudler, *Magical mushrooms, mystical molds* (Princeton, NJ: Princeton University Press, 1998); R. C. Moellering Jr. and G. M. Eliopoulos, *Principles of Anti-Infective Therapy* (Philadelphia: Elsevier, 2005), p. 242.

18. Hudler, *Magical mushrooms, mystical molds*; L. Kavalar, *Mushrooms, molds, and miracles* (New York: John Day Co., 1965).

19. G. F. Gensini et al., "The contributions of Paul Ehrlich to infectious disease," *Journal of Infection* 54 (2007): 221–24.

20. P. Ehrlich and S. Hata, *Die experimentelle chemotherapie der spirillosen* (Berlin: Julius Springer, 1910).

21. S. A. Waksman, "What is an antibiotic or an antibiotic substance," *Mycologia* 39 (1947): 565–69.

22. Ibid.; "Significant events of the last 125 years: 1933–1942," American Society for Microbiology, http://www.asm.org/MemberShip/index.asp?bid=17444 (accessed March 4, 2008). Note that, technically, sulfa drugs are not "antibiotics," since the definition Waksman originally came up with required the "antibiotic" compound to be produced by bacteria. In contrast, sulfa drugs are synthetic compounds manufactured by people. More accurately, sulfa drugs are referred to as "antibacterial agents" or "antimicrobial agents." But the difference is one of semantics, and outside of peer-reviewed scientific literature, sulfa drugs are often referred to by physicians as antibiotics.

23. S. A. Waksman, "Tenth anniversary of the discovery of streptomycin, the first chemotherapeutic agent found to be effective against tuberculosis in humans," *American Review of Tuberculosis* 70 (1954): 1–8.

24. A. Fleming, "On the antibacterial action of cultures of a *Penicillium* with special reference to their use in the isolation of *B. influenzae*," *British Journal*

of Experimental Pathology 10 (1929): 226–36; E. Lax, *The mold in Dr. Florey's coat: The story of the penicillin miracle* (New York: Henry Holt, 2004).

25. M. Wainwright and H. T. Swan, "C. G. Paine and the earliest surviving clinical records of penicillin therapy," *Medical History* 30 (1986): 42–56.

26. Lax, *The mold in Dr. Florey's coat*; E. Chain et al., "Penicillin as a chemotherapeutic agent," *Lancet* 2 (1940): 226–28; E. P. Abraham et al., "Further observations on penicillin," *Lancet* 2 (1941): 177–88.

27. R. Bud, "Antibiotics: The epitome of a wonder drug," *British Medical Journal* 334, Suppl. 1 (2007): S6.

28. "Funk and Wagnalls New Encyclopedia: DOMAGK, Gerhard Johannes Paul," World Almanac Education Group, http://www.history.com/encyclopedia.do?vendorId=FWNE.fw.do072850.a#FWNE.fw.do072850.a (accessed March 4, 2008); A. S. van Miert, "The sulfonamide-diaminopyrimidine story," *Journal of Veterinary Pharmacology and Therapeutics* 17 (1994): 309–16; L. Koss, "Book review: Gerhard Domagk, the first man to triumph over infectious diseases," *Human Pathology* 36 (2005): 1238–39.

29. E. H. Northey, *The sulfonamides and allied compounds* (New York: Reinhold Publishing, Inc., 1948).

30. Ibid.; S. H. Zinner and K. H. Mayer, *Sulfonamides and trimethoprim* (Philadelphia: Elsevier, 2005), p. 440.

31. Northey, *The sulfonamides and allied compounds*.

32. J. E. Lesch, *The first miracle drugs: How the sulfa drugs transformed medicine* (New York: Oxford University Press, 2007). Axillary temperatures are typically at least 1 degree Fahrenheit below oral temperatures. So, Hildegard Domagk's oral temperature at the start of prontosil therapy would likely have been in excess of 104 degrees Fahrenheit.

33. "Funk and Wagnalls New Encyclopedia: DOMAGK, Gerhard Johannes Paul"; "Significant events of the last 125 years: 1933–1942," American Society for Microbiology, http://www.asm.org/MemberShip/index.asp?bid=17444 (accessed March 4, 2008).

34. H. A. Carithers, "The first use of an antibiotic in America," *American Journal of Diseases of Children* 128 (1974): 207–11.

35. J. H. L. Heintzelman et al., "The use of p-aminobenzenesulphonamide in type 3 pneumococcus pneumonia," *American Journal of Medicine Science* 193 (1937): 759–63; G. M. Evans and W. F. Gaisford, "Treatment of pneumonia with 2-(p-aminobenzenesulphonamido) pyridine," *Lancet* 232, no. 5992 (1938): 14–19; A. L. Agranat et al., "Treatment of pneumonia with 2-(p-aminobenzene-

sulphonamido) pyridine," *Lancet* 233, no. 6024 (1939): 309–17; H. F. Flippin et al., "The treatment of pneumococcic pneumonia with sulfapyridine," *JAMA* 112 (1939): 529–34; D. Graham et al., "The treatment of pneumococcal pneumonia with dagenan (M. & B. 693)," *Canadian Medical Association Journal* 40 (1939): 325–32; B. W. Carey and T. B. Cooley, "Pneumonia in infants and children," *Journal of Pediatrics* 15 (1939): 613–21; W. F. Gaisford, "Results of the treatment of 400 cases of lobar pneumonia with M & B 693," *Proceedings of the Royal Society of Medicine* 32 (1939): 1070–76; T. Anderson and J. G. Cairns, "Treatment of pneumonia with sulphapyridine and serum," *Lancet* 236 (1940): 449–51; S. C. Wagoner and W. F. Hunting, "Sulfathiazole and sulfapyridine in the treatment of pneumonia in infancy and childhood," *JAMA* 116 (1941): 267–70; M. Finland, "Chemotherapy in bacteremias," *Connecticut State Medical Journal* 7 (1943): 92–100; G. A. H. Buttle, "The action of sulphanilamide and its derivatives with special reference to tropical diseases," *Transactions of the Royal Society of Tropical Medicine and Hygiene* 33 (1939): 141–58.

36. L. Thomas, *The youngest science: Notes of a medicine-watcher* (New York: Viking Press, 1983); M. Goldner, "Three generations of experience and thought in microbiology and infection," *Canadian Journal of Infectious Diseases* 14 (2003): 329–35.

37. B. Dixon, "Sulfa's true significance," *Microbes* 1 (2006): 500–501; W. M. Scheld and G. L. Mandell, "Sulfonamides and meningitis," *Journal of the American Medical Association* 251 (1984): 791–94.

38. B. Spellberg et al., "Position paper: Recommended design features of future clinical trials of anti-bacterial agents for community-acquired pneumonia," *Clinical Infectious Diseases* 47, Suppl. 3 (2008): S249–65.

39. Northey, *The sulfonamides and allied compounds.*

40. G. L. Mandell et al., ed., *Mandell, Douglas, and Bennett's principles and practice of infectious diseases* (Philadelphia: Churchill Livingstone, 2006).

41. R. Austrian and J. Gold, "Pneumococcal bacteremia with especial reference to bacteremic pneumococcal pneumonia," *Annals of Internal Medicine* 60 (1964): 759–76.

42. J. B. Bass Jr., "Lieutenant of the men of death," *Chest* 88 (1985): 483–84; M. Bliss, *William Osler: A life in medicine* (Toronto: University of Toronto Press, 1999).

43. Austrian and Gold, "Pneumococcal bacteremia with especial reference to bacteremic pneumococcal pneumonia."

44. K. A. Gordon et al., "Comparison of *Streptococcus pneumoniae* and

Haemophilus influenzae susceptibilities from community-acquired respiratory tract infections and hospitalized patients with pneumonia: Five-year results for the SENTRY antimicrobial surveillance program," *Diagnostic Microbiology and Infectious Disease* 46 (2003): 285–89; C. G. Whitney et al., "Increasing prevalence of multi-drug-resistant *Streptococcus pneumoniae* in the United States," *New England Journal of Medicine* 343 (2000): 1917–24; T. M. File Jr., "*Streptococcus pneumoniae* and community-acquired pneumonia: A cause for concern," *American Journal of Medicine* 117, Suppl. 3A (2004): 39S–50S; T. M. File Jr., "Clinical implications and treatment of multiresistant *Streptococcus pneumoniae* pneumonia," *Clinical Microbiology and Infection* 12, Suppl. 3 (2006): 31–41.

45. J. Turnidge and D. L. Paterson, "Setting and revising antibacterial susceptibility breakpoints," *Clinical Microbiology Reviews* 20 (2007): 391–408.

46. P. D. Brown et al., "Prevalence and predictors of trimethoprim-sulfamethoxazole resistance among uropathogenic *Escherichia coli* isolates in Michigan," *Clinical Infectious Diseases* 34 (2002): 1061–66; G. V. Doern, "The in vitro activity of cefotaxime versus bacteria involved in selected infections of hospitalized patients outside of the intensive care unit," *Diagnostic Microbiology and Infectious Disease* 22 (1995): 13–17; G. L. Drusano, "Role of pharmacokinetics in the outcome of infections," *Antimicrobial Agents and Chemotherapy* 32 (1988): 289–97; P. Gehanno et al., "In vivo correlates for *Streptococcus pneumoniae* penicillin resistance in acute otitis media," *Antimicrobial Agents and Chemotherapy* 39 (1995): 271–72; C. I. Kang et al., "Risk factors for and clinical outcomes of bloodstream infections caused by extended-spectrum beta-lactamase-producing *Klebsiella pneumoniae*," *Infection Control and Hospital Epidemiology* 25 (2004): 860–67; C. I. Kang et al., "Bloodstream infections due to extended-spectrum beta-lactamase-producing *Escherichia coli* and *Klebsiella pneumoniae*: Risk factors for mortality and treatment outcome, with special emphasis on antimicrobial therapy," *Antimicrobial Agents and Chemotherapy* 48 (2004): 4574–81; C. I. Kang et al., "Bloodstream infections caused by *Enterobacter* species: Predictors of 30-day mortality rate and impact of broad-spectrum cephalosporin resistance on outcome," *Clinical Infectious Diseases* 39 (2004): 812–18; J. D. Kellner et al., "Outcome of penicillin-nonsusceptible *Streptococcus pneumoniae* meningitis: A nested case-control study," *Pediatric Infectious Diseases Journal* 21 (2002): 903–10; C. Olivier et al., "Bacteriologic outcome of children with cefotaxime- or ceftriaxone-susceptible and -nonsusceptible *Streptococcus pneumoniae* meningitis," *Pediatric Infectious Diseases Journal* 19 (2000): 1015–17; R. Raz et al., "Empiric use of trimethoprim-sulfamethoxazole (TMP-SMX) in the treatment of women

with uncomplicated urinary tract infections, in a geographical area with a high prevalence of TMP-SMX–resistant uropathogens," *Clinical Infectious Diseases* 34 (2002): 1165–69; G. Sakoulas et al., "Relationship of MIC and bactericidal activity to efficacy of vancomycin for treatment of methicillin-resistant *Staphylococcus aureus* bacteremia," *Journal of Clinical Microbiology* 42 (2004): 2398–402; L. K. Hidayat et al., "High-dose vancomycin therapy for methicillin-resistant *Staphylococcus aureus* infections: Efficacy and toxicity," *Archives of Internal Medicine* 166 (2006): 2138–44; M. P. Pai et al., "Association of fluconazole area under the concentration-time curve/MIC and dose/MIC ratios with mortality in nonneutropenic patients with candidemia," *Antimicrobial Agents and Chemotherapy* 51 (2007): 35–39; J. J. Schentag et al., "*Streptococcus pneumoniae* bacteraemia: Pharmacodynamic correlations with outcome and macrolide resistance—a controlled study," *International Journal of Antimicrobial Agents* 30 (2007): 264–69.

47. "Bad bugs, no drugs: As antibiotic discovery stagnates, a public health crisis brews," Infectious Diseases Society of America (IDSA), http://www.idsociety.org/pa/IDSA_paper4_final_web.pdf (accessed April 30, 2005); "The Interagency Task Force on Antimicrobial Resistance and a public health action plan to combat antimicrobial resistance," Centers for Disease Control, http://www.cdc.gov/drugresistance/actionplan/index.htm (accessed December 1, 2008); S. R. Palumbi, "Humans as the world's greatest evolutionary force," *Science* 293 (2001): 1786–90; A. J. Alanis, "Resistance to antibiotics: Are we in the post-antibiotic era?" *Archives of Medical Research* 36 (2005): 697–705.

48. File Jr., "*Streptococcus pneumoniae* and community-acquired pneumonia: A cause for concern"; File Jr., "Clinical implications and treatment of multiresistant *Streptococcus pneumoniae* pneumonia."

49. S. Riedel et al., "Antimicrobial use in Europe and antimicrobial resistance in *Streptococcus pneumoniae*," *European Journal of Clinical Microbiology & Infectious Diseases* 26 (2007): 485–90.

50. M. E. Pichichero and J. R. Casey, "Emergence of a multiresistant serotype 19A pneumococcal strain not included in the 7-valent conjugate vaccine as an otopathogen in children," *JAMA* 298 (2007): 1772–78.

CHAPTER 3. METHICILLIN-RESISTANT *STAPHYLOCOCCUS AUREUS*— DEADLY ANTIBIOTIC-RESISTANT BACTERIA ESCAPE THE HOSPITAL

1. E. P. Abraham et al., "Further observations on penicillin," *Lancet* 2 (1941): 177–88.

2. Ibid.

3. W. Saxon, "Anne Miller, 90, first patient who was saved by penicillin," *New York Times*, http://www.wellesley.edu/Chemistry/Chem101/antibiotics/obit-a-miller.html (accessed March 4, 2008); C. M. Grossman, "The first use of penicillin in the United States," *Annals of Internal Medicine* 149 (2008): 135–36.

4. Grossman, "The first use of penicillin in the United States."

5. Saxon, "Anne Miller, 90, first patient who was saved by penicillin."

6. W. E. Herrell, "Further observations on the clinical use of penicillin," *Proceedings of the Staff Meetings of the Mayo Clinic* 18 (1943): 65–76; W. E. Herrell et al., "The clinical use of penicillin," *Proceedings of the Staff Meetings of the Mayo Clinic* 17 (1942): 609–16.

7. E. Lister, "Penicillin," *Lancet* 1 (1950): 645.

8. W. M. M. Kirby, "Extraction of a highly potent penicillin inactivator from penicillin resistant staphylococci," *Science* 99 (1944): 452–53; H. F. Chambers, "The changing epidemiology of *Staphylococcus aureus?*" *Emerging Infectious Diseases* 7 (2001): 178–82; A. A. Medeiros, "Evolution and dissemination of beta-lactamases accelerated by generations of beta-lactam antibiotics," *Clinical Infectious Diseases* 24, Suppl. 1 (1997): S19–45; C. H. Rammelkamp and T. Maxon, "Resistance of *Staphylococcus aureus* to the action of penicillin," *Proceedings of the Society for Experimental Biology and Medicine* 51 (1942): 386–89.

9. Chambers, "The changing epidemiology of *Staphylococcus aureus?*"; Medeiros, "Evolution and dissemination of beta-lactamases accelerated by generations of beta-lactam antibiotics"; S. K. Fridkin et al., "Methicillin-resistant *Staphylococcus aureus* disease in three communities," *New England Journal of Medicine* 352 (2005): 1436–44; M. Barber and J. E. M. Whitehead, "Bacteriophage types in penicillin-resistant staphylococcal infection," *British Medical Journal* 2 (1949): 565–69; R. I. Wise et al., "Personal reflections on nosocomial staphylococcal infections and the development of hospital surveillance," *Reviews in Infectious Diseases* 11 (1989): 1005–19.

10. Fridkin et al., "Methicillin-resistant *Staphylococcus aureus* disease in three communities"; H. F. Chambers, "Community-associated MRSA—Resistance and virulence converge," *New England Journal of Medicine* 352 (2005): 1485–87; L. G. Miller et al., "Necrotizing fasciitis caused by community-associated methicillin-resistant *Staphylococcus aureus* in Los Angeles," *New England Journal of Medicine* 352 (2005): 1445–53; G. J. Moran et al., "Methicillin-resistant *Staphylococcus aureus* in community-acquired skin infections," *Emerging Infectious Diseases* 11 (2005): 928–30; G. J. Moran et al., "Methicillin-resistant *S. aureus* infections among patients in the emergency department," *New England Journal of Medicine* 355 (2006): 666–74; E. A. Eady and J. H. Cove, "Staphylococcal resistance revisited: Community-acquired methicillin resistant *Staphylococcus aureus*—an emerging problem for the management of skin and soft tissue infections," *Current Opinions in Infectious Diseases* 16 (2003): 103–24; S. V. Kazakova et al., "A clone of methicillin-resistant *Staphylococcus aureus* among professional football players," *New England Journal of Medicine* 352 (2005): 468–75; C. A. Sattler et al., "Prospective comparison of risk factors and demographic and clinical characteristics of community-acquired, methicillin-resistant versus methicillin-susceptible *Staphylococcus aureus* infection in children," *Pediatric Infectious Diseases Journal* 21 (2002): 910–17.

11. R. M. Klevens et al., "Invasive methicillin-resistant *Staphylococcus aureus* infections in the United States," *JAMA* 298 (2007): 1763–71; E. A. Bancroft, "Antimicrobial resistance: it's not just for hospitals," *JAMA* 298 (2007): 1803–1804.

12. Chambers, "Community-associated MRSA—Resistance and virulence converge."

13. C. Liu et al., "A population-based study of the incidence and molecular epidemiology of methicillin-resistant *Staphylococcus aureus* disease in San Francisco, 2004–2005," *Clinical Infectious Diseases* 46 (2008): 1637–46.

14. G. L. Mandell et al., eds., *Mandell, Douglas, and Bennett's principles and practice of infectious diseases* (Philadelphia: Churchill Livingstone, 2006).

15. Miller et al., "Necrotizing fasciitis caused by community-associated methicillin-resistant *Staphylococcus aureus* in Los Angeles."

16. B. A. Diep et al., "The arginine catabolic mobile element and staphylococcal chromosomal cassette mec linkage: Convergence of virulence and resistance in the USA300 clone of methicillin-resistant *Staphylococcus aureus*," *Journal of Infectious Diseases* 197 (2008): 1523–30.

17. L. G. Miller and B. A. Diep, "Clinical practice: Colonization, fomites,

and virulence: Rethinking the pathogenesis of community-associated methicillin-resistant *Staphylococcus aureus* infection," *Clinical Infectious Diseases* 46 (2008): 752–60.

18. J. E. LaMar et al., "Sentinel cases of community-acquired methicillin-resistant *Staphylococcus aureus* onboard a naval ship," *Military Medicine* 168 (2003): 135–38.

19. M. W. Ellis et al., "Natural history of community-acquired methicillin-resistant *Staphylococcus aureus* colonization and infection in soldiers," *Clinical Infectious Diseases* 39 (2004): 971–79.

20. B. B. Pagac et al., "Skin lesions in barracks: Consider community-acquired methicillin-resistant *Staphylococcus aureus* infection instead of spider bites," *Military Medicine* 171 (2006): 830–32; F. J. Cloran, "Cutaneous infections with community-acquired MRSA in aviators," *Aviation, Space, and Environmental Medicine* 77 (2006): 1271–74.

21. D. M. Nguyen et al., "Recurring methicillin-resistant *Staphylococcus aureus* infections in a football team," *Emerging Infectious Diseases* 11 (2005): 526–32; A. J. Hall et al., "Multiclonal outbreak of methicillin-resistant *Staphylococcus aureus* infections on a collegiate football team," *Epidemiology and Infection* (2008): 1–9.

22. Kazakova et al., "A clone of methicillin-resistant *Staphylococcus aureus* among professional football players."

23. "Methicillin-resistant *Staphylococcus aureus* infections among competitive sports participants—Colorado, Indiana, Pennsylvania, and Los Angeles County, 2000–2003," *Morbidity and Mortality Weekly, CDC* 52 (2003): 793–95; "MRSA: Fighting the superbug," *60 Minutes*, CBS News, http://www.cbsnews.com/stories/2007/11/08/60minutes/main3474157.shtml (accessed November 12, 2007).

24. K. S. Kaye et al., "The deadly toll of invasive methicillin-resistant *Staphylococcus aureus* infection in community hospitals," *Clinical Infectious Diseases* 46 (2008): 1568–77.

25. J. E. Fergie and K. Purcell, "Community-acquired methicillin-resistant *Staphylococcus aureus* infections in south Texas children," *Pediatric Infectious Diseases Journal* 20 (2001): 860–63; B. C. Herold et al., "Community-acquired methicillin-resistant *Staphylococcus aureus* in children with no identified predisposing risk," *JAMA* 279 (1998): 593–98; M. J. Neff, "AAP, AAFP release guideline on diagnosis and management of acute otitis media," *American Family Physician* 69 (2004): 2713–15.

26. J. M. Lindenmayer et al., "Methicillin-resistant *Staphylococcus aureus* in a high school wrestling team and the surrounding community," *Archives of Internal Medicine* 158 (1998): 895–99.

27. Fridkin et al., "Methicillin-resistant *Staphylococcus aureus* disease in three communities"; Chambers, "Community-associated MRSA—Resistance and virulence converge"; Moran et al., "Methicillin-resistant *Staphylococcus aureus* in community-acquired skin infections"; Moran et al., "Methicillin-resistant *S. aureus* infections among patients in the emergency department"; W. J. Munckhof et al., "Emergence of community-acquired methicillin-resistant *Staphylococcus aureus* (MRSA) infection in Queensland, Australia," *International Journal of Infectious Diseases* 7 (2003): 259–64; K. Okuma et al., "Dissemination of new methicillin-resistant *Staphylococcus aureus* clones in the community," *Journal of Clinical Microbiology* 40 (2002): 4289–94.

28. M. Rao and T. Langmaid, "Bacteria that killed Virginia teen found in other schools," CNN, http://www.cnn.com/2007/HEALTH/10/18/mrsa.cases/index.html (accessed November 15, 2007); C. Cutright, "Teen dies of MRSA infection," *Roanoke Times*, http://www.roanoke.com/health/wb/136058 (accessed November 15, 2007).

29. Rao and Langmaid, "Bacteria that killed Virginia teen found in other schools"; "Staph infection worries close 21 Virginia schools," Reuters, http://www.reuters.com/article/domesticNews/idUSN1729913920071017 (accessed November 15, 2007).

30. L. Stahl, "MRSA: Fighting the superbug," *60 Minutes*, CBS News, http://www.cbsnews.com/stories/2007/11/08/60minutes/main3474157.shtml (accessed November 15, 2007).

31. "Local high school reports 5 football players with MRSA," NBC10.com, http://www.nbc10.com/health/14375221/detail.html?dl=mainclick (accessed November 15, 2007).

32. G. Barney, "MRSA outbreak causes scare," *Pittsburgh Post-Gazette Now*, http://www.post-gazette.com/pg/07355/3000000076.stm (accessed January 5, 2008).

33. "Six high school football players diagnosed with MRSA," WSOCTV.com, http://www.wsoctv.com/news/14367241/detail.html (accessed November 15, 2007).

34. "Strategies to address antimicrobial resistance act: Patient stories," Infectious Diseases Society of America, http://www.idsociety.org/STAARAct.htm (accessed October 10, 2007).

35. M. Baysallar et al., "Linezolid and quinupristin/dalfopristin resistance in vancomycin-resistant enterococci and methicillin-resistant *Staphylococcus aureus* prior to clinical use in Turkey," *International Journal of Antimicrobial Agents* 23 (2004): 510–12; L. Cui et al., "Correlation between reduced daptomycin susceptibility and vancomycin resistance in vancomycin-intermediate *Staphylococcus aureus*," *Antimicrobial Agents and Chemotherapy* 50 (2006): 1079–82; A. Mangili et al., "Daptomycin-resistant, methicillin-resistant *Staphylococcus aureus* bacteremia," *Clinical Infectious Diseases* 40 (2005): 1058–60; F. M. Marty et al., "Emergence of a clinical daptomycin-resistant *Staphylococcus aureus* isolate during treatment of methicillin-resistant *Staphylococcus aureus* bacteremia and osteomyelitis," *Journal of Clinical Microbiology* 44 (2006): 595–97; M. J. Peeters and J. C. Sarria, "Clinical characteristics of linezolid-resistant *Staphylococcus aureus* infections," *American Journal of Medicine Science* 330 (2005): 102–204; S. K. Pillai et al., "Linezolid resistance in *Staphylococcus aureus*: Characterization and stability of resistant phenotype," *Journal of Infectious Diseases* 186 (2002): 1603–2607; S. M. Roberts et al., "Linezolid-resistant *Staphylococcus aureus* in two pediatric patients receiving low-dose linezolid therapy," *Pediatric Infectious Diseases Journal* 25 (2006): 562–64; D. J. Skiest, "Treatment failure resulting from resistance of *Staphylococcus aureus* to daptomycin," *Journal of Clinical Microbiology* 44 (2006): 655–56; S. Tsiodras et al., "Linezolid resistance in a clinical isolate of *Staphylococcus aureus*," *Lancet* 358 (2001): 207–208; H. R. Vikram et al., "Clinical progression of methicillin-resistant *Staphylococcus aureus* vertebral osteomyelitis associated with reduced susceptibility to daptomycin," *Journal of Clinical Microbiology* 43 (2005): 5384–87; P. Wilson et al., "Linezolid resistance in clinical isolates of *Staphylococcus aureus*," *Journal of Antimicrobial Chemotherapy* 51 (2003): 186–88.

CHAPTER 4. BEYOND MRSA—INFECTIONS RESISTANT TO VIRTUALLY ALL ANTIBIOTICS

1. C. Dye, "Global epidemiology of tuberculosis," *Lancet* 367 (2006): 938–40.

2. "Trends in tuberculosis—United States, 2005," *MMWR Morbidity and Mortality Weekly Report* 55 (2006): 305–308.

3. G. R. Grant et al., "T. G. Heaton, tuberculosis, and artificial pneumothorax: Once again, back to the future?" *Chest* 112 (1997): 7–8; A. Kir et al.,

"Role of surgery in multi-drug-resistant tuberculosis: Results of 27 cases," *European Journal of Cardiothoracic Surgery* 12 (1997): 531–34; J. B. Nachega and R. E. Chaisson, "Tuberculosis drug resistance: A global threat," *Clinical Infectious Diseases* 36 (2003): S24–30; I. Strambu, "Therapeutic pneumothorax—An effective adjuvant method in treating multi-drug-resistant tuberculosis," *Pneumologia* 49 (2000): 129–36; S. W. Sung et al., "Surgery increased the chance of cure in multi-drug resistant pulmonary tuberculosis," *European Journal of Cardiothoracic Surgery* 16 (1999): 187–93; M. Zignol et al., "Global incidence of multi-drug-resistant tuberculosis," *Journal of Infectious Diseases* 194 (2006): 479–85; J. Zazueta-Beltran et al., "High rates of multi-drug-resistant *Mycobacterium tuberculosis* in Sinaloa State, Mexico," *Journal of Infection* 54 (2006): 411–12; "Multi-drug-resistant tuberculosis in Hmong refugees resettling from Thailand into the United States, 2004–2005," *MMWR Morbidity and Mortality Weekly Report* 54 (2005): 741–44; R. M. Granich et al., "Multi-drug-resistance among persons with tuberculosis in California, 1994–2003," *JAMA* 293 (2005): 2732–39; M. D. Nettleman, "Multi-drug-resistant tuberculosis: News from the front," *JAMA* 293 (2005): 2788–90.

4. "Extensively drug-resistant tuberculosis—United States, 1993–2006," *MMWR Morbidity and Mortality Weekly Report* 56 (2007): 250–53; A. Glusker, "Global tuberculosis levels plateau while extensively drug resistant strains increase," *British Medical Journal* 334 (2007): 659; H. Markel et al., "Extensively drug-resistant tuberculosis: An isolation order, public health powers, and a global crisis," *JAMA* 298 (2007): 83–86.

5. "Emergence of *Mycobacterium tuberculosis* with extensive resistance to second-line drugs—Worldwide, 2000–2004," *MMWR Morbidity and Mortality Weekly Report* 55 (2006): 301–305; N. R. Gandhi et al., "Extensively drug-resistant tuberculosis as a cause of death in patients co-infected with tuberculosis and HIV in a rural area of South Africa," *Lancet* 368 (2006): 1575–80; M. Raviglione, "XDR-TB: Entering the post-antibiotic era?" *International Journal of Tuberculosis and Lung Disease* 10 (2006): 1185–87.

6. F. Washburn, "Collapse therapy," *American Journal of Nursing* 37 (1937): 373–79.

7. I. Y. Motus et al., "Reviving an old idea: Can artificial pneumothorax play a role in the modern management of tuberculosis?" *International Journal of Tuberculosis and Lung Disease* 10 (2006): 571–77.

8. "Extensively drug-resistant tuberculosis—United States, 1993–2006"; Markel et al., "Extensively drug-resistant tuberculosis: An isolation order,

public health powers, and a global crisis"; R. Banerjee et al., "Extensively drug-resistant tuberculosis in California, 1993–2006," *Clinical Infectious Diseases* 47 (2008): 450–57.

9. "U.S. seeks fliers possibly exposed to rare TB," CNN, http://www.cnn.com/2007/HEALTH/conditions/05/29/tb.flight/ (accessed May 30, 2007); M. Stobbe, "TB case brings warning to air passengers," Associated Press, ABC News, http://abcnews.go.com/Health/wireStory?id=3222797 (accessed May 30, 2007); M. Stobbe, "Groom with TB under federal quarantine," Associated Press, ABC News, http://abcnews.go.com/Health/wireStory?id=3224668 (accessed May 30, 2007); G. Bluestein, "TB patient ID'd as Atlanta attorney, 31," Associated Press, ABC News, http://abcnews.go.com/US/wireStory?id=3231235 (accessed May 31, 2007).

10. "U.S. seeks fliers possibly exposed to rare TB."

11. Stobbe, "TB case brings warning to air passengers."

12. "U.S. health officials 'regretful' TB traveller fled Europe," CBC News, http://www.cbc.ca/world/story/2007/05/30/tb-flight.html (accessed June 10, 2007).

13. "News: Tuberculosis exposure feared on India-to-U.S. flight," *Clinical Infectious Diseases* 46 (2008): iii.

14. "Emergence of *Mycobacterium tuberculosis* with extensive resistance to second-line drugs—Worldwide, 2000–2004"; Gandhi et al., "Extensively drug-resistant tuberculosis as a cause of death in patients co-infected with tuberculosis and HIV in a rural area of South Africa"; Raviglione, "XDR-TB: Entering the post-antibiotic era?"

15. H. R. Kim et al., "Impact of extensive drug resistance on treatment outcomes in non-HIV-infected patients with multi-drug-resistant tuberculosis," *Clinical Infectious Diseases* 45 (2007): 1290–95; E. D. Chan et al., "Treatment outcomes in extensively resistant tuberculosis," *New England Journal of Medicine* 359 (2008): 657–59; M. C. Raviglione, "Facing extensively drug-resistant tuberculosis—A hope and a challenge," *New England Journal of Medicine* 359 (2008): 636–38.

16. L. C. McDonald, "Trends in antimicrobial resistance in healthcare-associated pathogens and effect on treatment," *Clinical Infectious Diseases* 42, Suppl. 2 (2006): S65–S71; J. E. McGowan Jr., "Resistance in nonfermenting gram-negative bacteria: Multi-drug-resistance to the maximum," *American Journal of Medicine* 119 (2006): S29–36, discussion S62–70; D. L. Paterson, "The epidemiological profile of infections with multi-drug-resistant *Pseudomonas aeruginosa*

and *Acinetobacter* species," *Clinical Infectious Diseases* 43, Suppl. 2 (2006): S43–48; P. R. Rhomberg et al., "Clonal occurrences of multi-drug-resistant Gram-negative bacilli: Report from the Meropenem Yearly Susceptibility Test Information Collection Surveillance Program in the United States (2004)," *Diagnostic Microbiology and Infectious Disease* 54 (2006): 249–57; S. H. Mirza et al., "Multi-drug resistant typhoid: A global problem," *Journal of Medical Microbiology* 44 (1996): 317–19; M. W. Douglas et al., "Multi-drug resistant *Pseudomonas aeruginosa* outbreak in a burns unit—An infection control study," *Burns* 27 (2001): 131–35; A. S. Levin et al., "Intravenous colistin as therapy for nosocomial infections caused by multi-drug-resistant *Pseudomonas aeruginosa* and *Acinetobacter baumannii*," *Clinical Infectious Diseases* 28 (1999): 1008–11; M. Muller et al., "Outbreaks of multi-drug resistant *Escherichia coli* in long-term care facilities in the Durham, York and Toronto regions of Ontario, 2000– 2002," *Canadian Communicable Disease Report* 28 (2002): 113–18; S. Zansky et al., "From the Centers for Disease Control and Prevention. Outbreak of multi-drug resistant *Salmonella* Newport—United States, January–April 2002," *JAMA* 288 (2002): 951–53; M. O. Wright, "Multi-resistant gram-negative organisms in Maryland: A statewide survey of resistant *Acinetobacter baumannii*," *American Journal of Infection Control* 33 (2005): 419–21; D. L. Paterson and Y. Doi, "A step closer to extreme drug resistance (XDR) in gram-negative bacilli," *Clinical Infectious Diseases* 45 (2007): 1179–81.

17. J. E. McGowan, "Resistance in nonfermenting gram-negative bacteria: Multi-drug-resistance to the maximum"; P. R. Rhomberg et al., "Clonal occurrences of multi-drug-resistant gram-negative bacilli: Report from the Meropenem Yearly Susceptibility Test Information Collection Surveillance Program in the United States (2004)"; B. Spellberg et al., "Trends in antimicrobial drug development: Implications for the future"; G. H. Talbot et al., "Bad bugs need drugs: An update on the development pipeline from the Antimicrobial Availability Task Force of the Infectious Diseases Society of America," *Clinical Infectious Diseases* 42 (2006): 657–68; M. M. Neuhauser et al., "Antibiotic resistance among gram-negative bacilli in United States intensive care units: Implications for fluoroquinolone use," *JAMA* 289 (2003): 885–88; T. R. Walsh, "The emergence and implications of metallo-beta-lactamases in gram-negative bacteria," *Clinical Microbiology and Infection* 11, Suppl. 6 (2005): 2–9.

18. T. Muratani and T. Matsumoto, "Urinary tract infection caused by fluoroquinolone- and cephem-resistant *Enterobacteriaceae*," *International Journal of Antimicrobial Agents* 28, Suppl. 1 (2006): S10–13.

19. Ibid.; G. G. Zhanel et al., "Antibiotic resistance in *Escherichia coli* outpatient urinary isolates: Final results from the North American Urinary Tract Infection Collaborative Alliance (NAUTICA)," *International Journal of Antimicrobial Agents* 27 (2006): 468–75; J. A. Karlowsky et al., "Fluoroquinolone-resistant urinary isolates of *Escherichia coli* from outpatients are frequently multidrug-resistant: Results from the North American Urinary Tract Infection Collaborative Alliance-Quinolone Resistance study," *Antimicrobial Agents and Chemotherapy* 50 (2006): 2251–54.

20. P. A. Bradford et al., "Emergence of carbapenem-resistant *Klebsiella* species possessing the class A carbapenem-hydrolyzing KPC-2 and inhibitor-resistant TEM-30 beta-lactamases in New York City," *Clinical Infectious Diseases* 39 (2004): 55–60; S. Bratu et al., "Rapid spread of carbapenem-resistant *Klebsiella pneumoniae* in New York City: A new threat to our antibiotic armamentarium," *Archives of Internal Medicine* 165 (2005): 1430–35; S. Bratu et al., "Emergence of KPC-possessing *Klebsiella pneumoniae* in Brooklyn, New York: Epidemiology and recommendations for detection," *Antimicrobial Agents and Chemotherapy* 49 (2005): 3018–20; S. Bratu et al., "Carbapenemase-producing *Klebsiella pneumoniae* in Brooklyn, NY: Molecular epidemiology and in vitro activity of polymyxin B and other agents," *Journal of Antimicrobial Chemotherapy* 56 (2005): 128–32; F. M. Kaczmarek et al., "High-level carbapenem resistance in a *Klebsiella pneumoniae* clinical isolate is due to the combination of bla(ACT-1) beta-lactamase production, porin OmpK35/36 insertional inactivation, and down-regulation of the phosphate transport porin phoe," *Antimicrobial Agents and Chemotherapy* 50 (2006): 3396–406.

21. McDonald, "Trends in antimicrobial resistance in healthcare-associated pathogens and effect on treatment"; McGowan, "Resistance in nonfermenting gram-negative bacteria: Multi-drug-resistance to the maximum"; Paterson, "The epidemiological profile of infections with multi-drug-resistant *Pseudomonas aeruginosa* and *Acinetobacter* species"; L. S. Munoz-Price and R. A. Weinstein, "*Acinetobacter* infection," *New England Journal of Medicine* 358 (2008): 1271–81; Walsh, "The emergence and implications of metallo-beta-lactamases in Gram-negative bacteria"; V. Aloush et al., "Multi-drug-resistant *Pseudomonas aeruginosa*: Risk factors and clinical impact," *Antimicrobial Agents and Chemotherapy* 50 (2006): 43–48; Y. Carmeli et al., "Health and economic outcomes of antibiotic resistance in *Pseudomonas aeruginosa*," *Archives of Internal Medicine* 159 (1999): 1127–32; H. Goossens, "Susceptibility of multi-drug-resistant *Pseudomonas aeruginosa* in intensive care units: Results from the European MYSTIC study

group," *Clinical Microbiology and Infection* 9 (2003): 980–83; A. Harris et al., "Epidemiology and clinical outcomes of patients with multiresistant *Pseudomonas aeruginosa*," *Clinical Infectious Diseases* 28 (1999): 1128–33; C. I. Kang et al., "*Pseudomonas aeruginosa* bacteremia: Risk factors for mortality and influence of delayed receipt of effective antimicrobial therapy on clinical outcome," *Clinical Infectious Diseases* 37 (2003): 745–51; M. D. Obritsch et al., "Nosocomial infections due to multi-drug-resistant *Pseudomonas aeruginosa*: Epidemiology and treatment options," *Pharmacotherapy* 25 (2005): 1353–64; F. Vidal et al., "Epidemiology and outcome of *Pseudomonas aeruginosa* bacteremia, with special emphasis on the influence of antibiotic treatment. Analysis of 189 episodes," *Archives of Internal Medicine* 156 (1996): 2121–26.

22. B. Spellberg et al., "Trends in antimicrobial drug development: Implications for the future," *Clinical Infectious Diseases* 38 (2004): 1279–86; G. H. Talbot et al., "Bad bugs need drugs: An update on the development pipeline from the Antimicrobial Availability Task Force of the Infectious Diseases Society of America," *Clinical Infectious Diseases* 42 (2006): 657–68.

23. A. J. Alanis, "Resistance to antibiotics: Are we in the post-antibiotic era?" *Archives of Medical Research* 36 (2005): 697–705; Raviglione, "XDR-TB: Entering the post-antibiotic era?"; M. E. Falagas and I. A. Bliziotis, "Pandrug-resistant gram-negative bacteria: The dawn of the post-antibiotic era?" *International Journal of Antimicrobial Agents* 29 (2007): 630–36; T. T. Yoshikawa, "Antimicrobial resistance and aging: Beginning of the end of the antibiotic era?" *Journal of the American Geriatric Society* 50 (2002): S226–29; J. W. Harrison and T. A. Svec, "The beginning of the end of the antibiotic era? Part I. The problem: Abuse of the 'miracle drugs,'" *Quintessence International* 29 (1998): 151–62; J. W. Harrison and T. A. Svec, "The beginning of the end of the antibiotic era? Part II. Proposed solutions to antibiotic abuse," *Quintessence International* 29 (1998): 223–29; R. E. W. Hancock, "The end of an era?" *Nature Reviews Drug Discovery* 6 (2007): 28.

24. Munoz-Price and Weinstein, "*Acinetobacter* infection"; S. Navon-Venezia et al., "High tigecycline resistance in multi-drug-resistant *Acinetobacter baumannii*," *Journal of Antimicrobial Chemotherapy* 59 (2007): 772–74.

25. "The Interagency Task Force on Antimicrobial Resistance and a public health action plan to combat antimicrobial resistance," Centers for Disease Control, http://www.cdc.gov/drugresistance/actionplan/index.htm (accessed December 1, 2008); B. Spellberg et al., "Societal costs versus savings from wild-card patent extension legislation to spur critically needed antibiotic develop-

ment," *Infection* 35 (2007): 167–74; E. Bouza et al., "*Pseudomonas aeruginosa*: A multicenter study in 136 hospitals in Spain," *Review Especial Quimioterapy* 16 (2003): 41–52; W. R. Jarvis and W. J. Martone, "Predominant pathogens in hospital infections," *Journal of Antimicrobial Chemotherapy* 29, Suppl. A (1992): 19–24.

26. McDonald, "Trends in antimicrobial resistance in healthcare-associated pathogens and effect on treatment"; McGowan, "Resistance in nonfermenting gram-negative bacteria: Multi-drug-resistance to the maximum"; T. M. File Jr., "*Streptococcus pneumoniae* and community-acquired pneumonia: A cause for concern," *American Journal of Medicine* 117, Suppl. 3A (2004): 39S–50S; T. M. File Jr., "Clinical implications and treatment of multiresistant *Streptococcus pneumoniae* pneumonia," *Clinical Microbiology and Infection* 12, Suppl. 3 (2006): 31–41; " The Interagency Task Force on Antimicrobial Resistance and a public health action plan to combat antimicrobial resistance"; A. J. Alanis, "Resistance to antibiotics: Are we in the post-antibiotic era?"; Granich et al., "Multi-drug-resistance among persons with tuberculosis in California, 1994–2003," *JAMA* 293 (2005): 2732–39; M. D. Nettleman, "Multi-drug-resistant tuberculosis: News from the front," *JAMA* 293 (2005): 2788–90; Y. Carmeli et al., "Health and economic outcomes of antibiotic resistance in *Pseudomonas aeruginosa*"; A. Harris et al., "Epidemiology and clinical outcomes of patients with multiresistant *Pseudomonas aeruginosa*"; "Antimicrobial resistance: Data to assess public health threat from resistant bacteria are limited. Report RCED-99–132," United States General Accounting Office, http://www.gao.gov/archive/1999/hx99132.pdf (accessed January 10, 2008); Brooklyn Antibiotic Resistance Task Force, "The cost of antibiotic resistance: Effect of resistance among *Staphylococcus aureus*, *Klebsiella pneumoniae*, *Acinetobacter baumannii*, and *Pseudmonas aeruginosa* on length of hospital stay," *Infection Control and Hospital Epidemiology* 23 (2002): 106–108; S. E. Cosgrove et al., "Health and economic outcomes of the emergence of third-generation cephalosporin resistance in *Enterobacter* species," *Archives of Internal Medicine* 162 (2002): 185–90; M. S. Niederman, "Impact of antibiotic resistance on clinical outcomes and the cost of care," *Critical Care Medicine* 29 (2001): N114–20.

27. M. S. Smolinski et al., *Microbial threats to health: Emergence, detection, and response* (Washington DC: Institute of Medicine, 2003).

28. "Bad bugs, no drugs: As antibiotic discovery stagnates, a public health crisis brews," Infectious Diseases Society of America (IDSA), http://www.idsociety.org/pa/IDSA_paper4_final_web.pdf (accessed April 30, 2005).

29. N. E. Aronson et al., "In harm's way: Infections in deployed American military forces," *Clinical Infectious Diseases* 43 (2006): 1045–51.

30. Ibid.; C. K. Murray et al., "Bacteriology of war wounds at the time of injury," *Military Medicine* 171 (2006): 826–29; D. E. Hinsley et al., "Ballistic fractures during the 2003 Gulf conflict—Early prognosis and high complication rate," *Journal of Royal Army Medical Corps* 152 (2006): 96–101.

31. Murray et al., "Bacteriology of war wounds at the time of injury."

32. J. Calhoun, "Extremity war injuries: State of the art & future directions. Session III: Antibiotics and war wounds," American Academy of Orthopedic Surgeons 2006 Annual Meeting, Chicago, IL (2006); K. M. Hujer et al., "Analysis of antibiotic resistance genes in multi-drug-resistant *Acinetobacter* sp. isolates from military and civilian patients treated at the Walter Reed Army Medical Center," *Antimicrobial Agents and Chemotherapy* 50 (2006): 4114–23; M. C. Albrecht et al., "Impact of *Acinetobacter* infection on the mortality of burn patients," *Journal of the American College of Surgeons* 203 (2006): 546–50.

33. D. R. Harman et al., "Aeromedical evacuations from Operation Iraqi Freedom: A descriptive study," *Military Medicine* 170 (2005): 521–27.

CHAPTER 5. LACK OF ANTIBIOTIC DEVELOPMENT

1. S. J. Projan and D. M. Shlaes, "Antibacterial drug discovery: Is it all downhill from here?" *Clinical Microbiology and Infection* 10, Suppl. 4 (2004): S18–22.

2. J. Travis, "Reviving the antibiotic miracle?" *Science* 264 (1994): 360–62.

3. D. M. Shlaes and R. C. Moellering Jr., "The United States Food and Drug Administration and the end of antibiotics," *Clinical Infectious Diseases* 34 (2002): 420–22.

4. D. N. Gilbert and J. E. Edwards Jr., "Is there hope for the prevention of future antimicrobial shortages?" *Clinical Infectious Diseases* 35 (2002): 215–16; author reply 16–17.

5. S. J. Projan, "Why is big pharma getting out of antibacterial drug discovery?" *Current Opinion in Microbiology* 6 (2003): 427–30.

6. Gilbert and Edwards, "Is there hope for the prevention of future antimicrobial shortages?"

7. B. Spellberg et al., "Trends in antimicrobial drug development: Implications for the future," *Clinical Infectious Diseases* 38 (2004): 1279–86.

8. M. S. Smolinski et al., *Microbial threats to health: Emergence, detection, and response* (Washington DC: Institute of Medicine, 2003); J. A. DiMasi et al., "The price of innovation: New estimates of drug development costs," *Journal of Health Economics* 22 (2003): 151–85; D. J. Payne et al., "Drugs for bad bugs: Confronting the challenges of antibacterial discovery," *Nature Reviews Drug Discovery* 6 (2007): 29–40; "New drug development: Science, business, regulatory, and intellectual property issues cited as hampering drug development efforts," US Government Accountability Office, http://oversight.house.gov/documents/20061219094529–73424.pdf (accessed September 21, 2007).

9. G. H. Talbot et al., "Bad bugs need drugs: An update on the development pipeline from the Antimicrobial Availability Task Force of the Infectious Diseases Society of America," *Clinical Infectious Diseases* 42 (2006): 657–68.

10. S. Gottlieb, "Attack of the superbugs," *Wall Street Journal*, http://aeipress.com/publications/filter.all,pubID.27037/pub_detail.asp (accessed March 3, 2008).

11. Spellberg et al., "Trends in antimicrobial drug development: Implications for the future."

12. "PhRMA '06–07 annual report," Pharmaceutical Research and Manufacturers of America, http://www.phrma.org/files/AR%202006–2007.pdf (accessed March 31, 2007); "Profile 2008: Pharmaceutical industry," Pharmaceutical Research and Manufacturers of America, www.phrma.org/files/2008%20Profile.pdf (accessed March 31, 2008).

13. J. P. Karlberg, "Trends in disease focus of drug development," *Nature Reviews Drug Discovery* 7 (2008): 639–40.

14. "Bad bugs, no drugs: As antibiotic discovery stagnates, a public health crisis brews," Infectious Diseases Society of America (IDSA), http://www.idsociety.org/pa/IDSA_paper4_final_web.pdf (accessed April 30, 2005).

15. M. Blaser, "The Honorable Richard Burr: Health, Education, Labor and Pensions Committee. Bioterrorism and Public Health Preparedness Subcommittee," http://72.14.253.104/search?q=cache:4ZVgUZ_i8N0J:www.idsociety.org/TemplateRedirect.cfm%3Ftemplate%3D/ContentManagement/ContentDisplay.cfm%26ContentID%3D16628+Letter+to+the+Honorable+Richard+Burr,+Chairman,+Subcommittee+on+Bioterrorism+and+Public+Health +Preparedness,+Senate+Committee+on+Health,+Education,+Labor+%26+Pensions&hl=en&gl=us&ct=clnk&cd=2&client=firefox-a (accessed November 21, 2006).

16. J. S. Bradley et al., "Anti-infective research and development: Problems, challenges, and solutions," *Lancet Infectious Diseases* 7 (2007): 68–78.

17. "Rx for survival: A global health challenge. Episode descriptions," WGBH/NOVA Science Unit and Vulcan Productions, Inc., http://www.pbs.org/wgbh/rxforsurvival/series/about/episodes.html (accessed January 18, 2008).

18. D. Bjerklie et al., "The year in medicine from A to Z: It was a year of old scourges and new drugs, from the first vaccine that prevents cancer to a bug that spoiled an entire crop of California spinach," *Time*, http://www.time.com/time/magazine/article/0,9171,1562958,00.html (accessed November 28, 2006).

19. "Strategies to Address Antimicrobial Resistance Act," Infectious Diseases Society of America, http://www.idsociety.org/STAARAct.htm (accessed October 10, 2007).

20. "IDSA endorses 50% tax credit for ID product research & development," Infectious Diseases Society of America, http://www.idsociety.org/WorkArea/showcontent.aspx?id=8450 (accessed November 26, 2007).

CHAPTER 6. THE WAR AGAINST MICROBES?

1. *Microbiology in the 21st century: Where are we and where are we going?* (Washington, DC: American Society for Microbiology, 2004).

2. Ibid.

3. R. D. Berg, "The indigenous gastrointestinal microflora," *Trends in Microbiology* 4 (1996): 430–35.

4. M. R. Fisk et al., "Alteration of oceanic volcanic glass: Textural evidence of microbial activity," *Science* 281 (1998): 978–80.

5. L. Perfeito et al., "Adaptive mutations in bacteria: High rate and small effects," *Science* 317 (2007): 813–15; J. W. Drake et al., "Rates of spontaneous mutation," *Genetics* 148 (1998): 1667–86.

6. D. Bensasson et al., "Genes without frontiers?" *Heredity* 92 (2004): 483–89.

7. Y. Doi and Y. Arakawa, "16S ribosomal RNA methylation: Emerging resistance mechanism against aminoglycosides," *Clinical Infectious Diseases* 45 (2007): 88–94.

8. W. Altermann and J. Kazmierczak, "Archean microfossils: A reappraisal of early life on earth," *Research in Microbiology* 154 (2003): 611–17; J. W. Schopf, "Microfossils of the Early Archean Apex chert: New evidence of the antiquity of life," *Science* 260 (1993): 640–46; J. W. Schopf and B. M. Packer,

"Early Archean (3.3-billion- to 3.5-billion-year-old) microfossils from Warrawoona Group, Australia," *Science* 237 (1987): 70–73.

9. B. Spellberg, "Bad bugs, no drugs: A failing police action (i.e., NOT a war)," 2004, Harold C. Neu Conference, Scottsdale, AZ; "Ending the war metaphor: The changing agenda for unraveling the host-microbe relationship—Workshop summary," National Academies Press, http://www.nap.edu/catalog/11699.html (accessed April 30, 2007).

10. S. J. Projan, "(Genome) size matters," *Antimicrobial Agents and Chemotherapy* 51 (2007): 1133–34.

11. A. A. Medeiros, "Evolution and dissemination of beta-lactamases accelerated by generations of beta-lactam antibiotics," *Clinical Infectious Diseases* 24, Suppl. 1 (1997): S19–45; B. G. Hall et al., "Independent origins of subgroup Bl + B2 and subgroup B3 metallo-beta-lactamases," *Journal of Molecular Evolution* 59 (2004): 133–41; B. G. Hall and M. Barlow, "Evolution of the serine beta-lactamases: Past, present and future," *Drug Resistance Update* 7 (2004): 111–23.

12. G. Dantas et al., "Bacteria subsisting on antibiotics," *Science* 320 (2008): 100–103.

13. J. Postgate, *Nitrogen Fixation* (Cambridge: Cambridge University Press, 1998).

14. M. J. Hill, "Intestinal flora and endogenous vitamin synthesis," *European Journal of Cancer Prevention* 6, Suppl. 1 (1997): S43–45.

15. S. R. Palumbi, "Humans as the world's greatest evolutionary force," *Science* 293 (2001): 1786–90.

16. "HHS Response to House Report 106–157—Agriculture, Rural Development, Food and Drug Administration, and Related Agencies, Appropriations Bill, 2000: Human-Use Antibiotics in Livestock Production," US Food and Drug Administration, Center for Veterinary Medicine, http://www.fda.gov/cvm/HRESP106_157.htm (accessed March 18, 2007); P. Collignon et al., "The routine use of antibiotics to promote animal growth does little to benefit protein undernutrition in the developing world," *Clinical Infectious Diseases* 41 (2005): 1007–13; S. Falkow and D. Kennedy, "Antibiotics, animals, and people—Again!" *Science* 291 (2001): 397; S. L. Gorbach, "Antimicrobial use in animal feed—Time to stop," *New England Journal of Medicine* 345 (2001): 1202–1203.

17. Palumbi, "Humans as the world's greatest evolutionary force"; S. Riedel et al., "Antimicrobial use in Europe and antimicrobial resistance in *Streptococcus pneumoniae*," *European Journal of Clinical Microbiology & Infectious Diseases*

26 (2007): 485–90; W. C. Albrich et al., "Antibiotic selection pressure and resistance in *Streptococcus pneumoniae* and *Streptococcus pyogenes*," *Emerging Infectious Diseases* 10 (2004): 514–17; H. Goossens et al., "Outpatient antibiotic use in Europe and association with resistance: A cross-national database study," *Lancet* 365 (2005): 579–87.

18. D. W. Rudge, "Myths about moths: A study in contrasts," *Endeavour* 30 (2006): 19–23; H. B. D. Kettlewell, *Industrial Melanism* (Oxford: Oxford University Press, 1973).

19. H. B. D. Kettlewell, "Further selection experiments on industrial melanism in the *Lepidoptera*," *Heredity* 10 (1956): 287–301.

20. Ibid; H. B. D. Kettlewell, "Selection experiments on industrial melanism in the *Lepidoptera*," *Heredity* 9 (1955): 323–42.

21. Rudge, "Myths about moths: A study in contrasts"; Kettlewell, *Industrial Melanism*.

22. Rudge, "Myths about moths: A study in contrasts"; Kettlewell, *Industrial Melanism*; Kettlewell, "Further selection experiments on industrial melanism in the *Lepidoptera*"; Kettlewell, "Selection experiments on industrial melanism in the *Lepidoptera*."

23. D. M. Shlaes and R. C. Moellering Jr., "The United States Food and Drug Administration and the end of antibiotics," *Clinical Infectious Diseases* 34 (2002): 420–22; T. F. O'Brien, "The global epidemic nature of antimicrobial resistance and the need to monitor and manage it locally," *Clinical Infectious Diseases* 24, Suppl. 1 (1997): S2–8; B. R. Levin et al., "The population genetics of antibiotic resistance," *Clinical Infectious Diseases* 24, Suppl. 1 (1997): S9–16; T. M. Barbosa and S. B. Levy, "The impact of antibiotic use on resistance development and persistence," *Drug Resistance Update* 3 (2000): 303–11.

24. S. K. Olofsson and O. Cars, "Optimizing drug exposure to minimize selection of antibiotic resistance," *Clinical Infectious Diseases* 45, Suppl. 2 (2007): S129–36.

25. S. Gottlieb, "Attack of the Superbugs," *Wall Street Journal*, http://aeipress.com/publications/filter.all,pubID.27037/pub_detail.asp (accessed March 3, 2008).

26. J. Lederberg, "Infectious history," *Science* 288 (2000): 287–93.

27. C. H. Jones, "The challenge of finding new antibacterials—Should we be hopeful?" Infectious Diseases Society of America (IDSA) Annual Meeting, San Diego, CA (2007).

28. S. J. Projan, "Why is big pharma getting out of antibacterial drug dis-

covery?" *Current Opinion in Microbiology* 6 (2003): 427–30; S. J. Projan, "New (and not so new) antibacterial targets—From where and when will the novel drugs come?" *Current Opinions in Pharmacology* 2 (2002): 513–22.

29. D. J. Payne et al., "Drugs for bad bugs: Confronting the challenges of antibacterial discovery," *Nature Reviews Drug Discovery* 6 (2007): 29–40; S. D. Mills, "When will the genomics investment pay off for antibacterial discovery?" *Biochemical Pharmacology* 71 (2006): 1096–1102; C. Leaf, "Why we're losing the war on cancer," *Fortune*, http://money.cnn.com/magazines/fortune/fortune_archive/2004/03/22/365076/index.htm (accessed March 31, 2007); G. Higgs, "Molecular genetics: The emperor's clothes of drug discovery?" *Drug Discovery Today* 9 (2004): 727–29; P. Balaram, "Drug discovery: Myth and reality," *Current Science* 87 (2004): 847–48; A. Aberg, "Bridging the gap between model systems and human biology," http://www.genengnews.com/articles/chitem.aspx?aid=878&chid=1 (accessed January 10, 2008); J. Rees, "Complex disease and the new clinical sciences," *Science* 296 (2002): 698–700; A. Abbott, "Cancer: The root of the problem," *Nature* 442 (2006): 742–43; M. Nagle, "Antisense drugs stop making sense? The once hyped antisense sector of the pharmaceutical industry has again proved a disappointment after one drug in development was refused approval and another was scrapped," DrugResearcher.com, http://www.drugresearcher.com/news/ng.asp?n=72869-genta-methylgene-antisense-mgi-pharma-isis (accessed March 27, 2007); P. Fernandes, "Antibacterial discovery and development—The failure of success?" *Nature Biotechnology* 24 (2006): 1497–503; A. F. Chalker and R. D. Lunsford, "Rational identification of new antibacterial drug targets that are essential for viability using a genomics-based approach," *Pharmacology & Therapeutics* 95 (2002): 1–20; C. Sheridan, "Antibiotics au naturel," *Nature Biotechnology* 24 (2006): 1494–96.

30. Payne et al., "Drugs for bad bugs: Confronting the challenges of antibacterial discovery"; "New drug development: Science, business, regulatory, and intellectual property issues cited as hampering drug development efforts," US Government Accountability Office, http://oversight.house.gov/documents/20061219094529–73424.pdf (accessed September 21, 2007); "Innovation or stagnation: Challenge and opportunity on the critical path to new medical products," US Department of Health and Human Services: Food and Drug Administration, http://www.fda.gov/oc/initiatives/criticalpath/whitepaper.html (accessed September 21, 2007); "FDA new molecular entities (NMEs) reports," US Food and Drug Administration, http://www.fda.gov/cder/rdmt/default.htm (accessed April 29, 2003); P. F. Chan et al., "Finding the gems using genomic

discovery: Antibacterial drug discovery strategies—The successes and the challenges," *Drug Discovery Today: Therapeutic Strategies* 1 (2004): 519–27.

31. J. A. DiMasi et al., "The price of innovation: New estimates of drug development costs," *Journal of Health Economics* 22 (2003): 151–85; "New drug development: Science, business, regulatory, and intellectual property issues cited as hampering drug development efforts"; "Innovation or stagnation: Challenge and opportunity on the critical path to new medical products."

32. "New drug development: Science, business, regulatory, and intellectual property issues cited as hampering drug development efforts"; "Innovation or stagnation: Challenge and opportunity on the critical path to new medical products."

33. S. J. Projan and D. M. Shlaes, "Antibacterial drug discovery: Is it all downhill from here?" *Clinical Microbiology and Infection* 10, Suppl. 4 (2004): S18–22; Projan, "Why is big pharma getting out of antibacterial drug discovery?"; Lederberg, "Infectious history"; Projan, "New (and not so new) antibacterial targets —From where and when will the novel drugs come?"; R. L. Monaghan and J. F. Barrett, "Antibacterial drug discovery—Then, now and the genomics future," *Biochemical Pharmacology* 71 (2006): 901–909; A. R. Coates and Y. Hu, "Novel approaches to developing new antibiotics for bacterial infections," *British Journal of Pharmacology* 152 (2007): 1147–54; K. M. Overbye and J. F. Barrett, "Antibiotics: Where did we go wrong?" *Drug Discovery Today* 10 (2005): 45–52.

34. DiMasi et al., "The price of innovation: New estimates of drug development costs"; Payne et al., "Drugs for bad bugs: Confronting the challenges of antibacterial discovery"; Mills, "When will the genomics investment pay off for antibacterial discovery?"; Chan et al., "Finding the gems using genomic discovery: Antibacterial drug discovery strategies—The successes and the challenges"; J. S. Blanchard, "Old approach yields new antibiotic," *Nature Medicine* 11 (2005): 1045–46.

35. Payne et al., "Drugs for bad bugs: Confronting the challenges of antibacterial discovery."

36. Jones, "The challenge of finding new antibacterials—Should we be hopeful?"

37. Payne et al., "Drugs for bad bugs: Confronting the challenges of antibacterial discovery"; Chan et al., "Finding the gems using genomic discovery: Antibacterial drug discovery strategies—The successes and the challenges."

38. DiMasi et al., "The price of innovation: New estimates of drug development costs."

39. Ibid.; "New drug development: Science, business, regulatory, and intellectual property issues cited as hampering drug development efforts."

40. "New drug development: Science, business, regulatory, and intellectual property issues cited as hampering drug development efforts."

41. Projan, "Why is big pharma getting out of antibacterial drug discovery?"; DiMasi et al., "The price of innovation: New estimates of drug development costs"; "New drug development: Science, business, regulatory, and intellectual property issues cited as hampering drug development efforts"; "FDA new molecular entities (NMEs) reports."

42. DiMasi et al., "The price of innovation: New estimates of drug development costs"; "Innovation or stagnation: Challenge and opportunity on the critical path to new medical products."

43. Spellberg et al., "Trends in antimicrobial drug development: Implications for the future."

44. Projan, "Why is big pharma getting out of antibacterial drug discovery?"

45. Spellberg et al., "Trends in antimicrobial drug development: Implications for the future."

46. "Guidelines for the use of antiretroviral agents in HIV-1-infected adults and adolescents," http://aidsinfo.nih.gov/contentfiles/Adultand AdolescentGL.pdf (accessed March 20, 2008).

47. J. Brown, "Approval of AZT," US Food and Drug Administration, http://www.fda.gov/bbs/topics/NEWS/NEW00217.html (accessed March 20, 2008).

48. "ACTG: AIDS Clinical Trial Group. About ACTG," ACTG, US Department of Health and Human Services, National Institute of Allergy and Infectious Diseases, Division of AIDS, http://www.aactg.org/node/1 (accessed December 17, 2008).

49. "ACTG: AIDS Clinical Trial Group Progress Report 2007–2008," ACTG, US Department of Health and Human Services, National Institute of Allergy and Infectious Diseases, Division of AIDS, http://www.aactg.org/sites/default/files/annual-progress-report-2007–2008.pdf (accessed December 17, 2008).

50. L. Montagnier, "Historical essay. A history of HIV discovery," *Science* 298 (2002): 1727–28.

51. R. C. Gallo et al., "Frequent detection and isolation of cytopathic retroviruses (HTLV-III) from patients with AIDS and at risk for AIDS," *Science* 224 (1984): 500–503; M. Popovic et al., "Detection, isolation, and continuous pro-

duction of cytopathic retroviruses (HTLV-III) from patients with AIDS and pre-AIDS," *Science* 224 (1984): 497–500.

52. J. A. Levy et al., "Isolation of lymphocytopathic retroviruses from San Francisco patients with AIDS," *Science* 225 (1984): 840–42.

53. Montagnier, "Historical essay. A history of HIV discovery."

54. A. Robicsek et al., "Universal surveillance for methicillin-resistant *Staphylococcus aureus* in 3 affiliated hospitals," *Annals of Internal Medicine* 148 (2008): 409–18.

55. L. B. Rice, "Federal funding for the study of antimicrobial resistance in nosocomial pathogens: No ESKAPE," *Journal of Infectious Diseases* 197 (2008): 1079–81.

56. T. H. Dellit et al., "Infectious Diseases Society of America and the Society for Healthcare Epidemiology of America guidelines for developing an institutional program to enhance antimicrobial stewardship," *Clinical Infectious Diseases* 44 (2007): 159–77.

57. Projan, "Why is big pharma getting out of antibacterial drug discovery?"; J. H. Powers, "Development of drugs for antimicrobial-resistant pathogens," *Current Opinions in Infectious Diseases* 16 (2003): 547–51.

58. M. J. Blaser and J. G. Bartlett, "Letter to FDA Commissioner Andrew C. von Eschenbach, MD," Infectious Diseases Society of America, http://www.idsociety.org/Template.cfm?Section=Home&CONTENTID=17039&TEMPLATE=/ContentManagement/ContentDisplay.cfm (accessed January 4, 2007); H. W. Boucher, "Open Public Forum," FDA Anti-Infective Drugs Advisory Committee: Joint with Drug Safety and Risk Management Advisory Committee, Silver Spring, MD.

59. J. Young et al., "Antibiotics for adults with clinically diagnosed acute rhinosinusitis: A meta-analysis of individual patient data," *Lancet* 371 (2008): 908–14.

60. "Clinical trial design for community-acquired pneumonia; Public workshop," March 1, 2008, http://www.fda.gov/cder/meeting/CAP.htm (accessed June 1, 2008).

61. B. Spellberg et al., "Position paper: Recommended design features of future clinical trials of anti-bacterial agents for community-acquired pneumonia," *Clinical Infectious Diseases* 47, Suppl. 3 (2008): S249–65.

62. Ibid.; M. Singer et al., "Historical and regulatory perspective on the treatment effect of antibacterial drugs in community-acquired pneumonia," *Clinical Infectious Diseases* 47, Suppl. 3 (2008): S216–24.

63. Spellberg et al., "Position paper: Recommended design features of future clinical trials of anti-bacterial agents for community-acquired pneumonia."

64. J. S. Bradley and G. H. J. McCracken, "Unique considerations for the evaluation of antibacterials in clinical trials of pediatric community-associated pneumonia," *Clinical Infectious Diseases* 47, Suppl. 3 (2008): S241–48.

65. W. R. Snodgrass and T. Anderson, "Prontosil in the treatment of erysipelas. A controlled series of 312 cases," *British Medical Journal* 2, no. 3933 (1937): 101–104; W. R. Snodgrass and T. Anderson, "Sulphanilamide in the treatment of erysipelas. A controlled series of 270 cases," *British Medical Journal* 2, no. 4014 (1937): 1156–59.

CHAPTER 7. FORGET PHARMACEUTICAL COMPANIES, THE GOVERNMENT CAN CREATE NEW ANTIBIOTICS—NOT!

1. K. Outterson et al., "Will longer antimicrobial patents improve global public health?" *Lancet Infectious Diseases* 7 (2007): 559–66.

2. B. Spellberg, "Antibiotic resistance and antibiotic development," *Lancet Infectious Diseases* 8 (2007): 211–12.

3. G. L. Mandell et al., eds., *Mandell, Douglas, and Bennett's principles and practice of infectious diseases* (Philadelphia: Churchill Livingstone, 2006).

4. P. G. Bray et al., "Quinolines and artemisinin: Chemistry, biology and history," *Current Topics in Microbiology and Immunology* 295 (2005): 3–38; Y. Tu, "The development of new antimalarial drugs: Qinghaosu and dihydroqinghaosu," *China Medical Journal (English)* 112 (1999): 976–77.

5. "WHO monograph on good agricultural and collection practices (GACP) for *Artemisia annua* L.," World Health Organization, http://www.who.int/medicines/publications/traditional/ArtemisiaMonograph.pdf (accessed March 25, 2007).

6. Bray et al., "Quinolines and artemisinin: Chemistry, biology and history"; D. Greenwood, "Conflicts of interest: The genesis of synthetic antimalarial agents in peace and war," *Journal of Antimicrobial Chemotherapy* 36 (1995): 857–72; L. W. Kitchen et al., "Role of US military research programs in the development of US Food and Drug Administration–approved antimalarial drugs," *Clinical Infectious Diseases* 43 (2006): 67–71.

7. G. A. H. Buttle, "The action of sulphanilamide and its derivatives with special reference to tropical diseases," *Transactions of the Royal Society of Tropical Medicine and Hygiene* 33 (1939): 141–58; Greenwood, "Conflicts of interest: The genesis of synthetic antimalarial agents in peace and war."

8. J. A. DiMasi et al., "The price of innovation: New estimates of drug development costs," *Journal of Health Economics* 22 (2003): 151–85.

9. "National Institutes of Health: Summary of the FY 2009 president's budget," National Institutes of Health Office of Budget, http://officeof budget.od.nih.gov/ui/HomePage.htm (accessed December 1, 2008).

10. J. A. DiMasi et al., "The price of innovation: New estimates of drug development costs"; C. Leaf, "Why we're losing the war on cancer," *Fortune*, http://money.cnn.com/magazines/fortune/fortune_archive/2004/03/22/365076/ index.htm (accessed March 31, 2007).

11. DiMasi et al., "The price of innovation: New estimates of drug development costs"; "Innovation or stagnation: Challenge and opportunity on the critical path to new medical products," US Department of Health and Human Services: Food and Drug Administration, http://www.fda.gov/oc/initiatives/ criticalpath/whitepaper.html (accessed September 21, 2007).

12. S. K. Fridkin et al., "Methicillin-resistant *Staphylococcus aureus* disease in three communities," *New England Journal of Medicine* 352 (2005): 1436–44; H. F. Chambers, "Community-associated MRSA—Resistance and virulence converge," *New England Journal of Medicine* 352 (2005): 1485–87; R. M. Klevens et al., "Invasive methicillin-resistant *Staphylococcus aureus* infections in the United States," *JAMA* 298 (2007): 1763–71; E. A. Bancroft, "Antimicrobial resistance: It's not just for hospitals," *JAMA* 298 (2007): 1803–1804; H. Wisplinghoff et al., "Nosocomial bloodstream infections in US hospitals: Analysis of 24,179 cases from a prospective nationwide surveillance study," *Clinical Infectious Diseases* 39 (2004): 309–17; L. S. Wilson et al., "The direct cost and incidence of systemic fungal infections," *Value Health* 5 (2002): 26–34; B. Spellberg et al., "Current treatment strategies for disseminated candidiasis," *Clinical Infectious Diseases* 42 (2006): 244–51.

13. B. J. Spellberg et al., "Efficacy of the anti-*Candida* rAls3p-N or rAls1p-N vaccines against disseminated and mucosal candidiasis," *Journal of Infectious Diseases* 194 (2006): 256–60; B. J. Spellberg et al., "The anti-*Candida albicans* vaccine composed of the recombinant N terminus of Als1p reduces fungal burden and improves survival in both immunocompetent and immunocompromised mice," *Infection & Immunity* 73 (2005): 6191–93; A. S. Ibrahim et al.,

"The anti-*Candida* rAls1p-N vaccine is broadly active against disseminated candidiasis," *Infection & Immunity* 74 (2006): 3039–41; A. S. Ibrahim et al., "Vaccination with recombinant N-terminal domain of Als1p improves survival during murine disseminated candidiasis by enhancing cell-mediated, not humoral, immunity," *Infection & Immunity* 73 (2005): 999–1005.

14. "Dale & Betty Bumpers Vaccine Research Center," US Department of Health and Human Services, National Institutes of Health, National Institute of Allergy and Infectious Diseases, http://www.vrc.nih.gov/ (accessed October 1, 2007).

15. "Dale & Betty Bumpers Vaccine Research Center: Mission statement," US Department of Health and Human Services, National Institutes of Health, National Institute of Allergy and Infectious Diseases, http://www.vrc.nih.gov/mission.htm (accessed October 1, 2007).

16. "NIAID Advanced Technology SBIR (NIAID-AT-SBIR [R43/R44])," US Department of Health and Human Services, http://grants.nih.gov/grants/guide/pa-files/PA-06–134.html (accessed October 1, 2007).

17. "SBIR/STTR Application Guide SF424 (R&R): A guide for preparing and submitting SBIR/STTR applications via grants.gov—Version 2," US Department of Health and Human Services: National Institutes of Health: Office of Extramural Research, available for download at http://grants.nih.gov/grants/funding/424/index.htm (accessed October 1, 2007).

18. S. J. Projan, "Why is big pharma getting out of antibacterial drug discovery?" *Current Opinion in Microbiology* 6 (2003): 427–30.

19. B. Spellberg et al., "Trends in antimicrobial drug development: Implications for the future," *Clinical Infectious Diseases* 38 (2004): 1279–86.

20. "Biopharmaceutical industry research & development tops $55 billion in 2006," Burrill & Company, http://www.burrillandco.com/burrill/pr_1171302475 (accessed March 31, 2007).

21. "Profile 2008: Pharmaceutical industry," Pharmaceutical Research and Manufacturers of America, www.phrma.org/files/2008%20Profile.pdf (accessed March 31, 2008).

22. "National Institutes of Health: Summary of the FY 2008 president's budget," US Department of Health and Human Services National Institutes of Health, http://officeofbudget.od.nih.gov/PDF/Press%20info-2008.pdf (accessed March 30, 2008).

23. H. Grabowski and J. Vernon, "Longer patents for increased generic competition in the United States. The Waxman-Hatch Act after one decade," *Pharmacoeconomics* 10, Suppl. 2 (1996): S110–23; R. J. Strongin, "Hatch-

Waxman, generics, and patents: Balancing prescription drug innovation, competition, and affordability," National Health Policy Forum background paper, http://www.nhpf.org/pdfs _bp/BP_HatchWaxman_6–02.pdf (accessed April 10, 2007); "FACT SHEET: Pharmaceutical patent incentives," PhRMA, http://www.phrma.org/publications/publications/17.06.2003.746.cfm (accessed June 15, 2005); M. E. Gluck, "Federal policies affecting the cost and availability of new pharmaceuticals," Georgetown University Institute for Healthcare Research and Policy and the Kaiser Family Foundation, http://www.kff.org/rxdrugs/loader=.cfm?url=/commonspot/security/getfile.cfm&PageID=14078 (accessed March 31, 2007).

24. *Fortune 500: How the industries stack up* (New York: Fortune Magazine, 2004), p. F26.

25. Ibid.

26. D. Fonda, "Inside the spore wars," Time, Inc., http://www.time.com/time/magazine/article/0,9171,1145253,00.html (accessed March 25, 2007).

27. Ibid.

28. M. Kaufman, "Bioterrorism response hampered by problem of profit," *Washington Post*, http://www.washingtonpost.com/wp-dyn/content/article/2005/08/06/AR2005080601164_pf.html (accessed March 23, 2007).

29. J. Guillemin, "Biological weapons and secrecy (WC 2300)," *FASEB Jour.* 19 (2005): 1763–65; M. Calabresi and M. August, "The smallpox scare," *Time*, http://www.time.com/time/magazine/article/0,9171,1101040726–665072,00.html (accessed April 3, 2008); D. J. Kuhles and D. M. Ackman, "The federal smallpox vaccination program: Where do we go from here?" *Health Affairs (Millwood)* Suppl. Web Exclusives (2003): W3–503–10.

30. Guillemin, "Biological weapons and secrecy (WC 2300)"; Calabresi and August, "The smallpox scare"; Kuhles and Ackman, "The federal smallpox vaccination program: Where do we go from here?"

31. A. S. Fauci, "Smallpox vaccination policy—The need for dialogue," *New England Journal of Medicine* 346 (2002): 1319–20.

32. Ibid.

33. Ibid.; "HHS awards $428 million contract to produce smallpox vaccine. Acambis/Baxter will produce 155 million doses by end of 2002," United States Department of Health and Human Services, http://www.hhs.gov/news/press/2001pres/20011128.html (accessed April 3, 2008).

34. L. DeFrancesco, "Throwing money at biodefense," *Nature Biotechnology* 22 (2004): 375–78; D. B. Resnik and K. A. DeVille, "Bioterrorism and patent

rights: "Compulsory licensure and the case of cipro," *American Journal of Bioethics* 2 (2002): 29–39.

35. M. S. Dworkin et al., "Fear of bioterrorism and implications for public health preparedness," *Emerging Infectious Diseases* 9 (2003): 503–505.

36. D. Shaffer et al., "Increased US prescription trends associated with the CDC *Bacillus anthracis* antimicrobial postexposure prophylaxis campaign," *Pharmacoepidemiology and Drug Safety* 12 (2003): 177–82; M. M'Ikanatha et al., "Patients' request for and emergency physicians' prescription of antimicrobial prophylaxis for anthrax during the 2001 bioterrorism-related outbreak," *BMC Public Health* 5 (2005): 2.

37. D. Fonda, "Inside the spore wars"; E. Bradley, "The worst-case scenario: Is America ready for a nuclear terrorist attack?" *60 Minutes*, CBS News, http://www.cbsnews.com/stories/2006/01/27/60minutes/main1245714.shtml (accessed March 23, 2007).

38. W. Swarts, "Government nukes Hollis-Eden's radiation drug," SmartMoney.com, http://www.smartmoney.com/onedaywonder/index.cfm?story=20070308&src=fb&nav=RSS20 (accessed June 30, 2007).

CHAPTER 8. "TOXIC PHARMACEUTICAL POLITICS" AND FINGER POINTING

1. "NIH budget remains flat in 2007," American Association for the Advancement of Science, http://www.aaas.org/spp/rd/nih07p.htm (accessed September 28, 2007).

2. D. G. Nathan and D. J. Weatherall, "Academic freedom in clinical research," *New England Journal of Medicine* 347 (2002): 1368–71.

3. Ibid.

4. Ibid.; J. Turk, "The greatest academic scandal of our era," Canadian Association of University Teachers, http://www.mcmaster.ca/mufa/turk.htm (accessed April 3, 2008).

5. N. Olivieri, "Movie review of *The Constant Gardener* by Dr. Nancy Olivieri," http://www.archivum.info/sci.med/2005–09/msg00127.html (accessed April 3, 2008).

6. L. P. Yang et al., "Deferasirox: A review of its use in the management of transfusional chronic iron overload," *Drugs* 67 (2007): 2211–30.

7. C. Reed et al., "Deferasirox, an iron-chelating agent, as salvage therapy for rhinocerebral mucormycosis," *Antimicrobial Agents and Chemotherapy* 50

(2006): 3968–69; A. S. Ibrahim et al., "The iron chelator deferasirox protects mice from mucormycosis through iron starvation," *Journal of Clinical Investigation* 117 (2007): 2649–57.

8. "Health expenditures," National Center for Health Statistics, http://www.cdc.gov/nchs/fastats/hexpense.htm (accessed March 27, 2007); "Breakdown of national healthcare expenditures in 2004," www.healthguideusa.org (accessed March 27, 2007); C. Smith et al., "National health spending in 2004: Recent slowdown led by prescription drug spending," *Health Affairs (Millwood)* 25 (2006): 186–96.

9. S. J. Projan, "Why is big pharma getting out of antibacterial drug discovery?" *Current Opinion in Microbiology* 6 (2003): 427–30.

10. C. Evans et al., "Use of quality adjusted life years and life years gained as benchmarks in economic evaluations: A critical appraisal," *Healthcare Management Science* 7 (2004): 43–49.

11. P. J. Neumann et al., "Are pharmaceuticals cost-effective? A review of the evidence," *Health Affairs (Millwood)* 19 (2000): 92–109.

12. J. Lynn and D. M. Adamson, "Living well at the end of life: Adapting healthcare to serious chronic illness in old age. Rand Health White Paper WP-137," Washington Home Center for Palliative Care Studies, http://www.medicaring.org/whitepaper/ (accessed March 27, 2007).

13. C. D. Baker et al., "Health of the nation—Coverage for all Americans," *New England Journal of Medicine* 359 (2008): 777–80.

14. E. J. Emanuel and V. R. Fuchs, "Who really pays for healthcare? The myth of 'shared responsibility,'" *JAMA* 299 (2008): 1057–59.

15. B. Spellberg et al., "Amphotericin B: Is a lipid-formulation gold standard feasible?" *Clinical Infectious Diseases* 38 (2004): 304–305; author reply, 306–307.

16. G. Hardin, "The tragedy of the commons," *Science* 162 (1968): 1243–48.

17. F. M. Scherer, "The link between gross profitability and pharmaceutical R&D spending," *Health Affairs (Millwood)* 20 (2001): 216–20.

18. R. J. Vogel, "Pharmaceutical patents and price controls," *Clinical Therapeutics* 24 (2002): 1204–22; discussion 1202–1203.

19. "Pharmaceutical price controls in OECD countries: Implications for U.S. consumers, pricing, research and development, and innovation," US Department of Commerce, International Trade Administration, http://www.ita.doc.gov/td/health/DrugPricingStudy.pdf (accessed April 7, 2007).

20. J. Reinhoudt, "Pharma in Europe: Going from heartburn to heart

attack?" American.com, http://www.american.com/archive/2007/january/pharma-in-europe-going-from-heartburn-to-heart-attack (accessed April 7, 2007); J. H. Golec and J. A. Vernon, "Euorpean pharmaceutical price regulation, firm profitability, and R&D spending. Working paper 12676," National Bureau of Economic Research, http://www.nber.org/papers/w12676 (accessed April 7, 2007) ; J. Chu, "How to plug Europe's brain drain," *Time*, http://www.time.com/time/printout/0,8816,901040119–574849,00.html (accessed April 7, 2007).

21. Reinhoudt, "Pharma in Europe: Going from heartburn to heart attack?"; Golec and Vernon, "European pharmaceutical price regulation, firm profitability, and R&D spending. Working paper 12676; Chu, "How to plug Europe's brain drain."

22. "Profile 2008: Pharmaceutical industry," Pharmaceutical Research and Manufacturers of America, www.phrma.org/files/2008%20Profile.pdf (accessed March 31, 2008).

23. Reinhoudt, "Pharma in Europe: Going from heartburn to heart attack?"; Golec and Vernon, "European pharmaceutical price regulation, firm profitability, and R&D spending. Working paper 12676; Chu, "How to plug Europe's brain drain."

24. Ibid.

25. "Would lower prescription drug prices curb drug company research & development?" Public Citizen, http://www.citizen.org/congress/reform/drug_industry/r_d/articles.cfm?ID=7909 (accessed April 7, 2007).

26. J. E. Calfee, "Pharmaceutical price controls and patient welfare," *Annals of Internal Medicine* 134 (2001): 1060–64.

27. Vogel, "Pharmaceutical patents and price controls"; "Pharmaceutical price controls in OECD countries: Implications for U.S. consumers, pricing, research and development, and innovation"; Golec and Vernon, "European pharmaceutical price regulation, firm profitability, and R&D spending. Working paper 12676"; R. J. Vogel, "Pharmaceutical pricing, price controls, and their effects on pharmaceutical sales and research and development expenditures in the European Union," *Clinical Therapeutics* 26 (2004): 1327–40; discussion 26; I. M. Cockburn, "The changing structure of the pharmaceutical industry," *Health Affairs (Millwood)* 23 (2004): 10–22; C. Giaccotto et al., "Drug prices and research and development investment behavior in the pharmaceutical industry," *Journal of Law and Economics* 48, pp. 195–214, http://www.journals.uchicago.edu/JLE/journal/issues/v48n1/480109/480109.web.pdf (accessed April 7, 2007); "Research and development in the pharmaceutical industry,"

Congress of the United States Congressional Budget Office (CBO), http://www.cbo.gov/ftpdocs/76xx/doc7615/10–02-DrugR-D.pdf (accessed April 7, 2007); J. A. Vernon, "Examining the link between price regulation and pharmaceutical R&D investment," *Health Economics* 14 (2005): 1–16.

28. T. Frankel, "Fiduciary law," *California Law Review* 71 (1983): 795; J. Diamond, *Collapse: How societies choose to fail or succeed*, chap. 15 (New York: Penguin, 2005), "Big Business and the Environment," pp. 483–84.

29. Spellberg et al., "Trends in antimicrobial drug development: Implications for the future," 38 (2004): 1279–86.

30. "FY 2008 budget," US Food and Drug Administration. Office of Management: Budget Formulation and Presentation, http://www.fda.gov/oc/oms/ofm/budget/2008/TOC.htm (accessed December 1, 2008).

31. "Strategies to address antimicrobial resistance act: Patient stories," Infectious Diseases Society of America, http://www.idsociety.org/STAARAct.htm (accessed October 10, 2007).

CHAPTER 9. SO WHAT WILL WORK?

1. "Bad bugs, no drugs: As antibiotic discovery stagnates, a public health crisis brews," Infectious Diseases Society of America (IDSA), http://www.idsociety.org/pa/IDSA_paper4_final_web.pdf (accessed April 30, 2005).

2. B. Spellberg et al., "The epidemic of antibiotic-resistant infections: A call to action for the medical community from the Infectious Diseases Society of America," *Clinical Infectious Diseases* 46 (2008): 155–64.

3. Ibid.

4. "United States Food and Drug Administration Office of Orphan Products Development," US Department of Health and Human Services, Food and Drug Administration, http://www.fda.gov/orphan/ (accessed September 21, 2007); "U.S. Food and Drug Administration definition of disease prevalence for therapies qualifying under the Orphan Drug Act," US Department of Health and Human Services: Food and Drug Administration, http://www.fda.gov/orphan/designat/prevalence.html (accessed September 21, 2007).

5. "United States Food and Drug Administration Office of Orphan Products Development."

6. "New drug development: Science, business, regulatory, and intellectual property issues cited as hampering drug development efforts," US Government

Accountability Office, http://oversight.house.gov/documents/20061219094529–73424.pdf (accessed September 21, 2007).

7. B. Spellberg, "Antibiotic resistance and antibiotic development," *Lancet Infectious Diseases* 8 (2008): 211–12.

8. M. Kaufman, "Bioterrorism response hampered by problem of profit," *Washington Post*, http://www.washingtonpost.com/wp-dyn/content/article/2005/08/06/AR2005080601164_pf.html (accessed March 23, 2007).

9. "GPhA: Legislation promoted as a countermeasure against bioterrorism would counter bipartisan measures to constrain prescription costs," Generic Pharmaceutical Association, http://www.prnewswire.com/cgi-bin/micro_stories.pl?ACCT=913120&TICK=GPHA&STORY=/www/story/02–08–2005/0002987118&EDATE=Feb+8,+2005 (accessed November 11, 2006).

10. "The Interagency Task Force on Antimicrobial Resistance and a public health action plan to combat antimicrobial resistance," Centers for Disease Control, http://www.cdc.gov/drugresistance/actionplan/index.htm (accessed December 1, 2008); "Antimicrobial resistance: Data to assess public health threat from resistant bacteria are limited. Report RCED-99–132," United States General Accounting Office, http://www.gao.gov/archive/1999/hx99132.pdf (accessed January 10, 2008).

11. B. Spellberg et al., "Societal costs versus savings from wild-card patent extension legislation to spur critically needed antibiotic development," *Infection* 35 (2007): 167–74.

12. Ibid.

CHAPTER 10. WHAT CAN YOU DO TO HELP?

1. A. C. Revkin, "Agency affirms human influence on climate," *New York Times*, http://www.nytimes.com/2007/01/10/science/10climate.html?ex1326085200&en=df1e7b40b876ac168 (accessed March 27, 2007).

2. "Bad bugs, no drugs: As antibiotic discovery stagnates, a public health crisis brews," Infectious Diseases Society of America (IDSA), http://www.idsociety.org/pa/IDSA_paper4_final_web.pdf (accessed April 30, 2005).

3. K. Outterson et al., "Will longer antimicrobial patents improve global public health?" *Lancet Infectious Diseases* 7 (2007): 559–66.

4. B. Spellberg, "Antibiotic resistance and antibiotic development," *Lancet Infectious Diseases* 8 (2008): 211–12.

5. K. Outterson, "Author's reply," *Lancet Infectious Diseases* 8 (2008): 212–13.

6. Outterson et al., "Will longer antimicrobial patents improve global public health?"

7. Ibid.

8. B. Spellberg et al., "Societal costs versus savings from wild-card patent extension legislation to spur critically needed antibiotic development," *Infection* 35 (2007): 167–74.

9. Spellberg, "Antibiotic resistance and antibiotic development."

10. Spellberg et al., "Societal costs versus savings from wild-card patent extension legislation to spur critically needed antibiotic development."

11. Outterson, "Author's reply."

12. K. L. McGinigle et al., "The use of active surveillance cultures in adult intensive care units to reduce methicillin-resistant *Staphylococcus aureus*-related morbidity, mortality, and costs: A systematic review," *Clinical Infectious Diseases* 46 (2008): 1717–25.

13. S. Harbarth et al., "Universal screening for methicillin-resistant *Staphylococcus aureus* at hospital admission and nosocomial infection in surgical patients," *JAMA* 299 (2008): 1149–57.

14. A. Robicsek et al., "Universal surveillance for methicillin-resistant *Staphylococcus aureus* in 3 affiliated hospitals," *Annals of Internal Medicine* 148 (2008): 409–18.

15. H. F. Chambers, "Community MRSA: Epidemiology and surveillance," in "Methicillin Resistant *Staphylococcus aureus* Initiative (MRSA-I): Addressing the challenges in the management and treatment of methicillin-resistant *Staphylococcus aureus*," American Society for Microbiology, www.mrsai.org (accessed January 30, 2009).

16. M. Edmond and T. C. Eickhoff, "Who is steering the ship? External influences on infection control programs," *Clinical Infectious Diseases* 46 (2008): 1746–50.

17. I. Wybo et al., "Outbreak of multi-drug-resistant *Acinetobacter baumannii* in a Belgian university hospital after transfer of patients from Greece," *Journal of Hospital Infection* 67 (2007): 374–80; Y. D. Podnos et al., "Eradication of multi-drug resistant *Acinetobacter* from an intensive care unit," *Surgical Infections* 2 (2001): 297–301.

18. A. D. Harris, "How important is the environment in the emergence of nosocomial antimicrobial-resistant bacteria?" *Clinical Infectious Diseases* 46

(2008): 686–88; E. Creamer and H. Humphreys, "The contribution of beds to healthcare-associated infection: The importance of adequate decontamination," *Journal of Hospital Infection* 69 (2008): 8–23.

19. Edmond and Eickhoff, "Who is steering the ship? External influences on infection control programs"; S. Saint et al., "Do physicians examine patients in contact isolation less frequently? A brief report," *American Journal of Infection Control* 31 (2003): 354–56; S. Tarzi et al., "Methicillin-resistant *Staphylococcus aureus*: Psychological impact of hospitalization and isolation in an older adult population," *Journal of Hospital Infection* 49 (2001): 250–54; H. T. Stelfox et al., "Safety of patients isolated for infection control," *JAMA* 290 (2003): 1899–905; H. E. Knowles, "The experience of infectious patients in isolation," *Nursing Times* 89 (1993): 53–56; K. B. Kirkland and J. M. Weinstein, "Adverse effects of contact isolation," *Lancet* 354 (1999): 1177–78; H. L. Evans et al., "Contact isolation in surgical patients: A barrier to care?" *Surgery* 134 (2003): 180–88; D. J. Diekema and M. B. Edmond, "Look before you leap: Active surveillance for multi-drug-resistant organisms," *Clinical Infectious Diseases* 44 (2007): 1101–1107; M. Z. Cohen et al., "Isolation in blood and marrow transplantation," *Western Journal of Nursing Research* 23 (2001): 592–609; G. Catalano et al., "Anxiety and depression in hospitalized patients in resistant organism isolation," *Southern Medical Journal* 96 (2003): 141–45.

20. Outterson, "Author's reply."

21. "National Institutes of Health: Summary of the FY 2008 president's budget," US Department of Health and Human Services National Institutes of Health, http://officeofbudget.od.nih.gov/PDF/Press%20info-2008.pdf (accessed March 30, 2008); "NIH budget remains flat in 2007," American Association for the Advancement of Science, http://www.aaas.org/spp/rd/nih07p.htm (accessed September 28, 2007).

CHAPTER 11. CONSEQUENCES AND CONCLUSIONS

1. L. Thomas, *The Youngest Science: Notes of a Medicine-Watcher* (New York: Viking Press, 1983).
2. Ibid., p. 13.
3. Ibid., p. 15.
4. Ibid., p. 16.
5. Ibid., pp. 19–20.
6. Ibid., p. 35.

index